RAND NATIONAL DEFENSE RESEARCH INSTITUTE

Capping Retired Pay for Senior Field Grade Officers

Force Management, Retention, and Cost Effects

Beth J. Asch, Michael G. Mattock, James Hosek, Patricia K. Tong

Prepared for the Office of the Secretary of Defense

For more information on this publication, visit www.rand.org/t/RR2251

Library of Congress Control Number: 2018937473

ISBN: 978-0-8330-9983-9

Support RAND

Make a tax-deductible charitable contribution at
www.rand.org/giving/contribute

www.rand.org

Preface

The Senate Armed Services Committee (SASC) directed the Secretary of Defense to conduct a study on the advisability and feasibility of reforming the 40-year pay table and retirement benefit to cap retired pay in certain ways. The committee expressed concern about findings from a 2016 U.S. Department of Defense (DoD) review that found that the most significant increases since 2007 in personnel with more than 30 years of service were not among general and flag officers and senior noncommissioned officers, but among field grade officers in the grades of O-4 to O-6, though the former group, not the latter, was the target of military compensation legislation in 2007 to increase incentives to serve for longer careers in the military. The SASC requested that the study consider the retention, cost, and force management effects of policies that cap retired pay based on the highest grade achieved, including a policy that would not allow officers with prior enlisted service to use their noncommissioned service to increase their retired pay.

The RAND National Defense Research Institute was asked to provide analytic support to DoD as input to its report to the SASC, and this report documents RAND's research. It should be of interest to policymakers and researchers concerned with military compensation and the relationship between the structure of the military pay table and the retention of enlisted and officer personnel.

This research was conducted within the Forces and Resources Policy Center of the RAND National Defense Research Institute, a federally funded research and development center sponsored by the Office of the Secretary of Defense, the Joint Staff, the Unified Combatant Commands, the Navy, the Marine Corps, the defense agencies, and the defense Intelligence Community.

For more information on the Forces and Resources Policy Center, see www.rand.org/nsrd/ndri/centers/frp or contact the director (contact information is provided on the webpage).

Contents

Figures

Tables

Summary

The Senate Armed Services Committee (SASC) report accompanying the fiscal year (FY) 2017 National Defense Authorization Act (NDAA; Pub. Law 114-328) directed the Secretary of Defense to review the advisability and feasibility of reforming the military basic pay table and retirement benefit to cap retired pay by pay grade so that only members of the highest grades and with the most years of service (YOS) would earn the highest retirement benefits.[1] The motivation for the directive was a significant increase since 2007 in the number of field grade officers in the grades of O-4 to O-6 with more than 30 YOS reported in a 2016 U.S. Department of Defense (DoD) review of the military basic pay table. While Title 10 of the U.S. Code restricts service beyond 30 years for active component officers in pay grades O-4 to O-6 who are not on promotion lists, the restrictions can be waived, and the restrictions are based on years of commissioned service, not including prior service as an enlisted member.[2]

The 2016 review was supported by analysis presented in Asch et al. (2016). The 2007 NDAA (Pub. Law 109-364) extended the military basic pay table to 40 YOS, lifted a cap on the retired pay of senior military personnel, and made additional compensation changes to increase the incentives for the most senior personnel, especially general and flag officers and senior noncommissioned officers, to stay in service longer. The retention of field grade officers beyond 30 YOS was not the intended target of the 2007 NDAA.

The 2016 SASC review directed that the Secretary of Defense review reforms that would cap retired pay so that only members of the highest grades and with the most YOS would earn the highest retirement benefits. The study should

- include an assessment of the cost savings, impact on morale and retention, promotion rates, and force management
- consider cost-saving measures that still allow members with 20 YOS to retire but prevent officers with prior enlisted service from using noncommissioned time served to increase their retired pay
- consider the suitability of special and incentive (S&I) pays as an alternative retention tool to the increased retired pay multiplier created by the FY 2007 NDAA, to compensate

---

[1] Throughout the report, *YOS* refers to years of active component service, including both enlisted service and officer service for officers who have enlisted service before becoming an officer. We will distinctly state years of commissioned service when we mean commissioned service only.

[2] See 10 U.S.C. Sections 631–636 and Section 661. These rules were created under the Defense Officer Personnel Management Act (DOPMA) (Pub. Law 96-513). DOPMA pertains to the Air Force, Army, Marine Corps, and Navy. The Coast Guard normally operates under the U.S. Department of Homeland Security and is governed by a different set of rules. The statutes governing Coast Guard officers are in 14 U.S.C. Chapter 11 (Kapp, 2016).

specific occupational specialties that have limited promotion rates but greater longevity benefits, such as chaplains and limited duty officers.

The research summarized in this report provides analysis to support the Secretary of Defense review. Our study took a multi-method approach that focused on active duty personnel and drew on expert knowledge, administrative data for retention tabulations, and advanced econometric methods for estimating retention behavior and for simulating policy responses. We used data from the Defense Manpower Data Center on active duty personnel by grade, YOS, and service to tabulate personnel strength and retention of officers with more than 30 YOS from 2000 to 2016 to better understand the extent to which observed increases in field grade officer strength before and after 2007 are attributable to increases in the number and retention of officers with prior enlisted service. In addition, we conducted 14 interviews of DoD civilian and military experts who have insight into compensation and the management of senior military personnel to qualitatively assess the effects of capping retired pay on retention, morale, promotion rates, and force management. We also used RAND's Dynamic Retention Model (DRM) capability to simulate the active component retention and cost effects of capping retired pay by grade and of using S&I pays as an alternative retention tool to increasing the retired pay multiplier. The DRM is an econometric model of individual retention behavior in the military that was used to support the analysis in the 2016 RAND report by Asch et al., as well as analyses of the retention and cost effects of other changes to military compensation, including pay raises and retirement reform proposals. Because our suite of existing DRMs does not focus on the retention behavior of officers with prior enlisted service, as part of this study we extended the DRM to model the retention behavior of officers with prior enlisted service, constructed data files that longitudinally track the individual careers of officers with and without prior enlisted service, and used the files to estimate Army and Navy DRMs of officers with and without prior enlisted service. We used these estimated models to simulate the retention and cost effects of retired pay caps.

Key Findings

The Number of Senior Field Grade Officers with Prior Enlisted Service Has Increased

Our tabulations of the number of officers by grade and YOS between 2000 and 2016 show that the increase in the number of officers with more than 30 YOS reported in the 2016 study was attributable to an increase in officers with prior enlisted service. That number increased since 2000 while the number without prior enlisted service decreased, though between 2007 and 2013 the number without prior enlisted service was relatively stable. The net effect was an increase in the total number with more than 30 YOS beginning in 2007, thereby explaining the result found in the 2016 study that the number of officers with more than 30 YOS increased markedly after 2007. We found that nearly all of the officers with more than 30 YOS in the grades of O-4 and O-5 and the majority in the grade of O-6 are officers with prior enlisted service. Among the services, the Army and Navy employ the greatest number of officers with prior enlisted service with more than 30 YOS, and the growth in the number with prior enlisted service in these two services has been quite dramatic.

The experts we interviewed provided several explanations for the increase in the number of field grade officers with prior enlisted service. Many mentioned an increased requirement for longer careers among those with specialized skills and knowledge, and described officers with enlisted experience as being a key source of technical expertise.[3] Others mentioned that the increase could be a residual effect of commissioning enlisted personnel to meet officer short-falls in the 1990s and early 2000s. Since up-or-out rules for officers after 30 YOS are based on commissioned service, officers with prior enlisted service are not required to retire under these rules when they have 30 YOS, since some of their YOS are enlisted years.

Capping Retired Pay Would Reduce Retention of Officers Who Come from the Enlisted Ranks

The general view from the experts we interviewed was that capping retired pay along the lines suggested by the SASC would hurt officer retention and morale and, possibly, the accession of officers who come up from the enlisted ranks. Our simulations of the retention effects of cap-ping retired pay using the newly estimated DRMs for Army and Navy officers supports this conclusion. We found that preventing the use of prior enlisted service in the retired pay multi-plier computation significantly changed the experience mix of the force, with far fewer officers with prior enlisted service staying until 20 YOS, where YOS included both officer and enlisted years. The result is fewer person-years of service before 20 YOS. But, among those who do stay to 20 YOS, far more stay beyond 20 YOS. Thus, the seniority of this group of Army and Navy officers would increase, but fewer would choose such a long career.

We also simulated the effect of capping the retired pay multiplier for officers in grades O-5 and below so that their retired pay multiplier did not increase beyond 30 YOS, meaning additional years beyond 30 for these officers would not increase the multiplier. We compare the retention of officers when the cap is changed with retention under a baseline where the cap is defined by current policy and specifically where the cap does not vary with grade. There is no presumption that future requirements will call for the retention profile and experience mix produced under the baseline under current policy, but, short of knowing what future experi-ence mix requirements are, the baseline of retention under current policy serves as a useful benchmark for assessing retention effects of policy changes. We found that retention of officers with and without prior enlisted service would decline, especially beyond 30 YOS, but the effect was much larger for officers with prior enlisted service. DOPMA rules affect officers with no prior enlisted service at 30 YOS but affect officers with prior enlisted service after 30 YOS.

Sustaining Retention Using Special and Incentive Pays Would Result in a Net Increase in Cost

One approach to managing the lower retention of officers is to offer S&I pay to sustain reten-tion. In both types of multiplier caps we examined—not counting years of prior enlisted ser-vice, and not counting years beyond 30 for officers in grades O-5 and below—using S&I pays to restore overall retention would result in a net cost increase.

Our DRM simulations indicate that, when prior enlisted service YOS are not counted toward retirement, more, rather than fewer, members with prior enlisted service would choose to serve beyond 30 years absent any S&I pay. Any simple retention incentive pay would only

---

[3]   We conducted 14 interviews, so we do not have a large enough sample size to make meaningful quantitative inferences about the preponderance of responses. We indicate when only one or two interviewees gave a given response. When we say *few* or *several*, we mean that three to five interviewees gave the response. When we say *many*, we mean the majority gave the response.

serve to exacerbate the problem and increase cost. Restoring the baseline force profile of members with prior enlisted service would require a combination of retention incentive pays in early YOS and separation incentive pays in later YOS targeted solely at members with prior enlisted service—a policy that would be difficult to defend in practice.

If the retirement pay multiplier were instead capped at 30 YOS for officers in grades O-5 and below, our simulations indicate that using S&I pay to restore retention would result in a 0.5 to 0.6 percent increase in net cost to the services.

Interviews Indicated That Capping Retired Pay Is Not Advisable

The experts we interviewed did not consider S&I pay a feasible or desirable alternative to retired pay, though S&I pay could have a positive retention effect. A particular concern among some interviewees was that even with the addition of S&I pay, a cap on the retired pay multiplier could be perceived by service members as a cut to military compensation, especially on top of the recent military retirement reform, and negative perceptions could hurt retention, thereby increasing the S&I pay—and cost—of sustaining retention. Another concern they expressed was that S&I pay is subject to uncertainty, so retired pay was considered potentially more valuable than S&I pay.

Many of the experts also agreed that the extended pay table and current retired pay cap had a number of advantages for force management and readiness and so should be retained. They stated that DoD needs the flexibility to keep some personnel for long careers, and so it is important that the incentives are in place to encourage them to serve longer. The view was that there is a continued need to have flexibility to manage the force and retain people with critical skills that are costly and time-consuming to develop, and that the extended pay table and current retired pay cap provide such flexibility. Many interviewees said it was important to maintain the flexibility provided by the ability to access and retain officers with prior enlisted service. These officers are an important source of technical expertise, and a higher return on their training is achieved when they stay for long careers. They also provide a flexible and ready source of accessions during times of officer shortages. More broadly, the experts argued that it was important to consider the long-term horizon when contemplating changes to military compensation and cautioned against reforming compensation in such a way that could be detrimental to future readiness. Put differently, nearly all stated that capping retired pay was not advisable.

Conclusion

The number of officers with more than 30 years total of enlisted soldier and commissioned officer experience has increased since 2000. While the 2007 reform of the military pay table provides incentives for service beyond 30 years, the increase in officers with more than 30 YOS and prior enlisted service predates the 2007 reform. As one of the interviewees observed, the increased retention of these officers was unexpected but not unwelcomed.

Capping retirement pay multipliers seems a reasonable approach to try to control personnel costs for these officers; however, in our DRM simulations of two alternative capping approaches, we found that once measures have been taken to restore retention, little savings remain on the table. Furthermore, the subject-matter experts we interviewed on the matter feared the deleterious effect that measures to cap the retirement pay multiplier might have on morale, and expressed concern about the practicality of using, and justifying the use, of S&I pays to sustain retention if caps were put in place. Therefore, we conclude that capping retirement pay for senior field grade officers would have no positive benefits and could result in harmful outcomes with respect to force management and cost.

Acknowledgments

We are indebted to the military personnel management and compensation experts who participated in the interviews we conducted. We are grateful to Don Svendsen in the Office of Compensation within the Office of the Under Secretary of Defense for Personnel and Readiness, who served as project monitor, provided background material, and arranged interviews within the project's tight timeline. We are also grateful to Jeri Busch, director of that office, who provided input and guidance to our analysis. Our study and report also benefited from the peer reviews we received from Ellen Pint at RAND and from Amalia Miller at the University of Virginia. At RAND, we wish to thank Arthur Bullock and Anthony Lawrence, who created data files and ran the retention tabulations in Chapter Two, and Sean McKenna for research assistance.

CHAPTER ONE

Introduction

The motivation for this study was a directive by the Senate Armed Services Committee (SASC) in its report accompanying the fiscal year (FY) 2017 National Defense Authorization Act (NDAA; Pub. Law 114-328) for the Secretary of Defense to review the advisability and feasibility of reforming the military basic pay table and retirement benefit to cap retired pay by pay grade. The objective of the new reforms would be to allow only members in the highest grades, not the mid grades, and with the most years of service (YOS) to earn the highest retirement benefits.[1] That is, members in lower grades but with more YOS would have capped retired pay, while members in higher grades with more YOS would have uncapped retired pay. Basic pay is the foundation of military compensation and is determined by a set of tables for commissioned officers, warrant officers, and enlisted personnel that show how pay varies with YOS and pay grade. Pay is higher for those in higher grades, and, within a grade, pay increases as a result of longevity increases.

Four compensation changes were made in the FY 2007 NDAA (Pub. Law 109-364) to provide larger incentives for the most experienced members, particularly general and flag officers, to continue to serve and to reward such service. First, until 2007, longevity pay increases in basic pay stopped at YOS 26. This pay structure has been termed the *30-year basic pay table*; members serving beyond 26 years no longer received basic pay increases as a result of additional seniority. The FY 2007 NDAA created the so-called 40-year pay table; it added longevity pay increases after YOS 26 for officers in pay grades O-6 and above, warrant officers in pay grades W-4 and W-5, and enlisted members in pay grades E-8 and E-9. Second, the 2007 legislation eliminated the cap on the multiplier of basic pay for the purpose of computing retired pay. Under the 30-year pay table, the multiplier was limited to 75 percent at YOS 30, but, because the cap was removed under the 2007 reforms, the multiplier for members who served beyond 30 years continued to increase.[2] The change in the multiplier was not conditioned on grade, so that the multiplier increased with YOS regardless of the member's pay grade. Third, the legislation raised the cap on basic pay for general and flag officers (O-7 to O-10) from Execu-

---

[1]   Throughout the report, *YOS* refers to years of active component service, including both enlisted service and officer service for officers who have enlisted service before becoming an officer. We will distinctly state years of commissioned service when we mean commissioned service only.

[2]   Under the Blended Retirement System (BRS) introduced by the 2016 NDAA (Pub. Law 114-92), the multiplier would be 80 percent of basic pay for a service member with 40 YOS. New entrants, as of January 1, 2018, will be automatically covered by BRS, while currently serving members are grandfathered under the legacy retirement systems. Those with 12 or fewer YOS as of December 31, 2017, will be permitted to opt into BRS. Because of this timeline, members with 30 or more YOS will only be under BRS beginning in 2036.

tive Schedule Level III to Executive Schedule Level II, and fourth, it removed the Executive Schedule Level cap on basic pay for the purpose of computing retired pay.[3]

The 2017 SASC-requested review is a follow-on to a study it requested in 2015 to review the military pay tables, focusing on whether the 40-year pay table was still justified as a retention tool. The SASC report accompanying the 2015 NDAA questioned whether it was useful to continue the 40-year table from a retention standpoint, or whether the retention of experienced personnel who would otherwise be difficult to retain could be achieved with a 30-year pay table.

The Secretary of Defense provided a report in response to the 2015 SASC request in which it concluded that the 40-year pay table was still justified as a retention tool. The report drew on RAND analysis (Asch et al., 2016) showing that the number of active duty personnel with more than 30 YOS increased by nearly 60 percent after 2007, though increases began even before 2007. The greatest percentage increase was not among general and flag officers, the group that represented the impetus for the 2007 legislative changes, but was among enlisted personnel, particularly E-9s and field grade officers in pay grades O-4 to O-6. The latter group increased by nearly 50 percent. Because O-6s are not permitted to serve more than 30 years of commissioned service unless they have a waiver, it was hypothesized that the increases in field grade officers were mostly among officers with prior enlisted service. The study found that continuation rates beyond 26 YOS did not increase markedly overall after 2007, despite the increase in the number of personnel with more than 30 YOS. These findings imply that the increase in the number of personnel with more than 30 YOS came from an increase in the size of the cohorts reaching 26 YOS. This suggests that requirements for senior personnel may have increased after 2007 and, if so, that these requirements were filled by senior personnel in very specific groups, such as officers with prior enlisted experience who stayed for an extra assignment and recalled retirees who returned to support the increased pace of deployment.

Asch et al. (2016) found that a 30-year table could be as effective at sustaining retention as a 40-year table, as long as the services had adequate special pay to manage retention of senior personnel, and the cost would be $1.2 billion per year less (in 2014 dollars) than the cost of keeping the 40-year table. That said, the study also found, based on interviews with senior personnel managers, that the 40-year pay table performed well, with many arguing that it improved readiness and flexible personnel management. The majority of interviewees stated that reverting to the 30-year table could adversely affect morale and perceptions about the stability and value of military compensation overall.

For the 2017 follow-on study, the SASC report directed the Secretary of Defense to review reforms that would cap retired pay so that only members of the highest grades and with the most YOS would earn the highest retirement benefits, with separate caps on retired pay for commissioned and noncommissioned officers (NCOs), warrant officers, and enlisted personnel. It directed that the study should

- include an assessment of the cost savings, impact on morale and retention, promotion rates, and force management
- consider cost-saving measures that still allow members with 20 YOS to retire but prevent officers with prior enlisted service from using noncommissioned time served to increase their retired pay

---

[3]   This cap was reinstated in the FY 2015 NDAA (Pub. Law 113-291).

- consider the suitability of special and incentive (S&I) pays as an alternative retention tool to the increased retired pay multiplier created by the FY 2007 NDAA, to compensate specific occupational specialties that have limited promotion rates but greater longevity benefits, such as chaplains and limited duty officers.

The SASC directive did not provide an explicit reason for capping retired pay, but the directive's language included discussion of the 2016 finding that the most significant percentage increases in service over 30 years were among field grade officers, a finding that seems counter to the stated purpose of extending the 2007 compensation changes to induce longer careers, especially among general and flag officers. The SASC-proposed cap on retired pay would involve reducing the incentives of field grade officers to serve for long careers. The objective appears to be to change the grade mix of the longest-serving officers, so that those with the most YOS would be primarily those in the highest grades. The SASC language also focused on limiting longer service among officers with prior enlisted service, by considering a cap that would prevent the inclusion of enlisted service for the purposes of increasing retired pay. On the other hand, the language recognized that longer service is desirable in some specialties and communities where promotion is limited, such as chaplains and limited duty officers, and requested an analysis of the use of S&I pays as an alternative to retired pay to induce longer careers for these personnel.

The directive required the study to consider the effects of capping retired pay on morale, retention, promotion, force management, and cost. Capping retired pay for senior field grade officers could be viewed as a cut in compensation and could hurt morale. The cap could hurt retention, not only among highly senior field grade officers but also among more-junior officers who are forward-looking and might see the expected value of staying in the military as less valuable. It could potentially affect the willingness of enlisted personnel to join the officer corps. On the other hand, promotion opportunities for more-junior officers might improve if retention falls among more-senior field grade officers. The effects on force management will depend on the services' requirements for experience among field grade officers, especially experience obtained during prior enlisted service, and how the services manage their officer corps and the role of officers with prior enlisted service in meeting accessions. Personnel costs might also be affected by a cap on retired pay. Such a cap could reduce retirement costs associated with those whose retired pay is capped. But retirement costs could increase, as could basic pay costs, if the services substitute higher-grade general and flag officers for lower-grade field grade officers to meet their requirements for experience, when retired pay for lower-grade personnel is capped. Costs could also increase if retention is sustained by offering S&I pay. Thus, while the SASC directive presumed there would be a cost savings to capping retired pay, the net effect on cost is unclear, a priori, and requires modeling and analysis to estimate. Our analysis provides information on the direction of the effect on cost.

The research summarized in this report provides analysis to meet the objectives of the directive and to support the Secretary of Defense review in response to this follow-on 2017 SASC request. The approach draws on expert knowledge, administrative data for retention tabulations, and advanced econometric methods for estimating retention behavior and for simulating responses to policy changes.

Specifically, we conducted interviews of U.S. Department of Defense (DoD) civilian and military experts who have insight into compensation and the management of senior military personnel. These interviews provided background information on the possible reasons for the observed increases in the number of field grade officers and NCOs with more than 30 YOS found in the 2016 RAND study, as well as qualitative assessments of the effects of capping retired pay on retention, morale, promotion rates, and force management. In addition, we used active duty pay files from the Defense Manpower Data Center (DMDC) to update the tabulations on the personnel strength and retention of personnel with more than 30 YOS presented in the 2016 RAND study. Because SASC requested that the study consider measures that would limit the ability of officers with prior enlisted service to use noncommissioned time served to increase retired pay, we also used the DMDC data to tabulate the personnel strength and retention rates of officers with and without prior enlisted service from 2000 to 2016. These tabulations provide context on the extent to which observed increases in field grade officer strength before and after 2007 are attributable to increases in the number and retention of officers with prior enlisted service.[4]

We used RAND's Dynamic Retention Model (DRM) capability to simulate the retention and cost effects of capping retired pay by grade and of using S&I pays as an alternative retention tool to increasing the retired pay multiplier. The DRM is an econometric model of individual retention behavior in the military that has been documented extensively and has been used for analyses of the retention and cost effects of other changes to military compensation, including pay raises and retirement reform proposals. The DRMs for officers estimated for these earlier applications focus on the retention behavior of officers with no prior enlisted service, not that of officers with prior enlisted service. Furthermore, for the most part, the models did not permit analysis of the retention and cost effects of compensation policies that differ by grade, such as different retired pay caps by grade. Thus, as part of this study, we constructed data files that longitudinally track the individual careers of officers with prior enlisted service and used the files to estimate an Army and Navy DRM of officers with prior enlisted service, extending the DRM to allow for prior enlisted service as well as promotion and grade. While an early version of the DRM included grade (Asch et al., 2008), later versions omitted it, because adding grade as another "state" that defined a member's career status at a point in time increased the computational burden, yet the policies that were the focus of our efforts, namely military retirement reform, were not grade- or promotion-specific, so incorporating grade was not needed. Because the SASC request involves analysis of grade-specific policies (capping the retirement pay of field grade officers), extending the DRM again to include grade and promotion was necessary. We used the new model estimates to simulate retention and cost effects of retired pay caps.

The remainder of this report is organized as follows. Chapter Two presents the tabulations of personnel with more than 30 YOS, while Chapter Three summarizes the main themes that emerged from the interviews we conducted. The extension of the DRM to officers with prior enlisted service and to allow for promotion, including data development and model estimates, is shown in Chapter Four. Chapter Five summarizes the key findings from simulations conducted with the DRM. We present our conclusions in Chapter Six. Appendix A provides some additional tabulations by service, and Appendix B provides the protocol used for interviews with DoD subject-matter experts.

---

[4]   This study underwent RAND Institutional Review Board review and was deemed to be research not involving human subjects.

CHAPTER TWO

Updated Trends in the Number of Officers with More Than 30 Years of Service

In our 2016 report, we tabulated the number of personnel who remained on active duty past 30 YOS, from 2000 to 2014.[1] In this chapter, we update these tabulations through 2016, and we further decompose the tabulations for officers to consider trends in the number of officers with and without prior enlisted service. As in the previous study, we used active duty pay files from the DMDC and specifically the September inventory of personnel from 2000 through 2016, including only active duty personnel. We created cross-tabulations by YOS; by pay grade and YOS; and by pay grade, service, and YOS. We computed YOS using the pay entry base date.[2] We conclude the chapter with tabulations of the retention profiles of officers with and without prior enlisted service and the distribution of years of prior enlisted service among those officers with prior enlisted service.

Results from Our Previous Report

We found that the overall number of personnel with more than 30 YOS had increased since 2007, especially in the Army but also in the other services. Although there were differences across services, these increases were greatest among senior enlisted personnel, specifically E-9s, and warrant officers. For officers, the greatest increase in personnel with more than 30 YOS was among O-5s and O-6s, possibly individuals with prior enlisted service or recalled retirees, as opposed to flag grade officers (O-7 to O-10). In addition, the upward trends that we observed in the number of personnel with more than 30 YOS often begin well before, and, in some cases, continued well after, 2007 and may be tied to the increased demands placed on the military during the Afghanistan and Iraq wars, the global recession, and other contextual factors. We also examined continuation rates among personnel with more than 30 YOS, but did not find a significant increase in retention rates for senior personnel after 2007. This was true across services and pay grades, with a few exceptions.

The lack of change in continuation rates, coupled with an observed increase in the number of personnel serving past 30 years, suggest that those increases in personnel with more than 30 YOS may have been concentrated among very specific groups of people, such as officers

---

[1] Asch et al. (2016) also computed continuation rates for personnel with more than 26 YOS but found little change in them.

[2] Pay entry base date is adjusted for breaks in service and includes periods of service during which a member is entitled to retired pay (DoD, 2017).

with prior enlisted experience who stayed for an extra assignment, senior enlisted and warrant officers who similarly were retained to fill specific jobs, and recalled retirees who returned to support the increased pace of deployment. It could also be attributable in part to an increase in the overall size of the military in the period since 2001, driven by increased demands and pace of deployment in this period, rather than an increase in the percentage of personnel choosing to stay in the military past 30 years. We did not explore these alternative explanations in the previous report, nor do we do so in this report.

Updated Results on the Number of Officers with More Than 30 Years of Service

One hypothesis for the observed increase in the number of personnel with more than 30 YOS after 2007, especially among field grade rather than general and flag officers, is a possible increase in the number of field grade officers with prior enlisted service. The 1980 Defense Officer Personnel Management Act (DOPMA; Pub. Law 96-513) limits the number of years of commissioned service an officer may serve before being mandatorily or voluntarily retired.[3] Officers in the grade of O-5 face mandatory retirement if they are not on the promotion list to O-6 after 28 years of commissioned service, while officers in the grade of O-6 not on a promotion list to O-7 face mandatory retirement after 30 years of commissioned service (Title 10 U.S.C., Sections 633 and 634). The exception is for certain officers in the Navy and Marine Corps who are either limited duty officers or permanent professors at the U.S. Naval Academy. Officers with prior enlisted service must have at least 10 years of commissioned service to retire as an officer, but because these personnel become officers only after serving in the enlisted ranks, they may not reach 28 years of commissioned service as an O-5 or 30 years of commissioned service as an O-6 until they have more than 30 years of total active service.

The number of officers with prior enlisted service may have increased for several reasons. First, with the legislative changes that occurred in 2007, especially the changes in the retirement calculations that allowed service after 30 years to count in the retirement formula, these individuals had an increased incentive to stay beyond 30 years after 2007. However, it is also the case that they had an increased incentive to leave the military to begin claiming the higher retirement benefits.

Second, the overall demand for field grade officers with more than 30 YOS may have increased due to the operational requirements associated with Afghanistan and Iraq. Such demand could have been met by officers with prior enlisted service and with officers with no prior enlisted service but who were permitted by the secretary of their service to continue beyond the mandatory retirement point because of the needs of the service. Department of Defense Instruction 1320.08 was issued in March 2007, just one month before the adoption of the new 40-year basic pay table in April 2007, to provide policy guidelines to the services for the continuation of commissioned officers on the active duty and reserve duty retirement lists. The instruction requires the services to convene continuation selection boards based on the needs of the service for the continuation of officers on these lists who have specific needed

---

[3]   See 10 U.S.C. Sections 631–636 and Section 661. DOPMA pertains to the Air Force, Army, Marine Corps, and Navy. The Coast Guard normally operates under the U.S. Department of Homeland Security and is governed by a different set of rules. The statutes governing Coast Guard officers are in 14 U.S.C. Chapter 11 (Kapp, 2016).

skills or qualifications. In addition, during time of war or national emergency, the services can recall retirees involuntarily, or retirees can volunteer for Retiree Recall if they are qualified. For example, Army Regulation 601-10 describes the management and recall of Army retirees in support of mobilization and peacetime operations. Thus, members who would otherwise be required to retire may receive a waiver to serve beyond 30 YOS if there is need by the service.

Third, the proportion of officers with prior enlisted service may have increased because of retention issues among officers with no prior service. That is, the services may have had an increase in demand for junior officers in general, but the demand was met with an increase in the number of officers with prior enlisted service because of insufficient retention among officers without prior enlisted service. Over time, these junior officers with prior enlisted service continue in service and get promoted, reaching the field grades later in their careers.

The tabulations below provide additional information on the trends in the number of officers with more than 30 YOS by decomposing the number into those with and without prior enlisted service. By considering prior enlisted status, we can examine how much of the increase in the number of officers serving after 2007 is attributable to an increase in the number of officers with prior enlisted service. The interviews we conducted with DoD personnel and manpower experts and discussed in Chapter Three provide additional insights into these hypotheses.

Updated Data

As we did in the 2016 report, we use active duty pay files, but we update the data through FY 2016. In addition, for officers, we merge the officer data with enlisted pay data to identify officers who have prior enlisted service.

Total Number of Officers with More Than 30 Years of Service

Our tabulations show that the number of officers with more than 30 YOS increased since FY 2007 (Figure 2.1).[4] Beyond 2014, the year our earlier studied ended, we find that the number stabilized in 2015 but continued to increase in 2016. The figure also shows the trend over time among officers with more than 30 YOS with and without prior enlisted service. The number of prior enlisted officers in this group has steadily increased since 2000, from 650 officers in 2000 to 1,466 in 2007, an increase of 126 percent. Between 2007 and 2016, the number further increased to 3,020, an increase of 106 percent. In contrast, the number of officers with no prior enlisted service began to decline after 2002 from 1,971 in 2002 to 1,380 in 2007. After 2007, the number without prior enlisted service was relatively stable at around 1,550 until 2011, when it began to decline again to 1,175 by 2016.

The trends show that the number of officers with more than 30 YOS and with prior enlisted service has indeed increased while the number without prior enlisted service has decreased, but these trends began before the 2007 legislative changes to the pay table and retirement formula. Prior to 2007, the increase in the number with prior enlisted service offset the decrease in the number without prior enlisted service, so the total was relatively stable. After 2007, the number with prior enlisted service continued to increase while the number without prior enlisted service stabilized, so the total increased. Beyond 2011, when the number

---

4   In the remainder of this report, all reference to years are to fiscal years.

Figure 2.1
Number of Active Duty Officers with More Than 30 Years of Service, Total and by Whether Officer Has Prior Enlisted Service

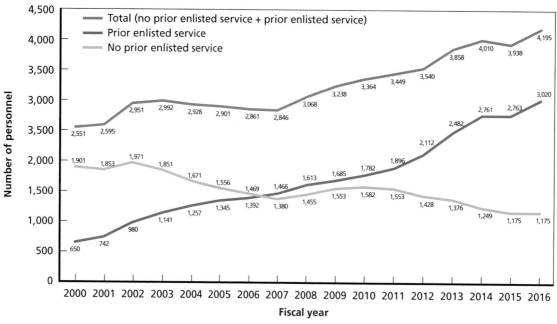

without prior enlisted service began to decrease again, the decline was more than offset by the increase in the number of officers with prior enlisted service, so the total continued to increase.

It is useful to put the number of officers with more than 30 YOS in context relative to the overall number of officers by examining the percentage of all officers (O-4 and above) who have more than 30 YOS. Figure 2.2 shows that, across DoD, while the percentage has increased since 2000, it is at most 5 percent. Thus, the overall percentage of officers with more than 30 YOS is quite small. The percentage is highest in the Army and Navy, reaching 6.1 percent and 6.7 percent, respectively, in 2016, and lowest in the Air Force. The percentage with more than 30 YOS varies by pay grade (Figure 2.3). Nearly 70 percent of general and flag officers have more than 30 YOS in 2016, whereas less than 15 percent of O-6s and almost 6 percent of O-5s have more than 30 YOS. The percentage of O-4s with more than 30 YOS is less than 1 percent.

Tabulations by Grade

Figures 2.4 to 2.7 show the number of officers with more than 30 YOS in grades O-4, O-5, O-6, and O-7 through O-10, respectively. Relatively few officers in the grade of O-4 serve beyond 30 YOS, but those who do are primarily officers with prior enlisted service, especially in more recent years. The number of O-4 officers with more than 30 YOS with prior enlisted service steadily grew between 2000 and 2014, but the number has declined since then. Between 2000 and 2007, the number with prior enlisted service grew by 116 percent, from 62 to 134 officers, and grew further between 2007 and 2014 to 412 officers, an increase of 207 percent. The number fell to 354, or by 14 percent, by 2016.

Figure 2.2
Percentage of Active Duty Officers with More Than 30 Years of Service, by Service

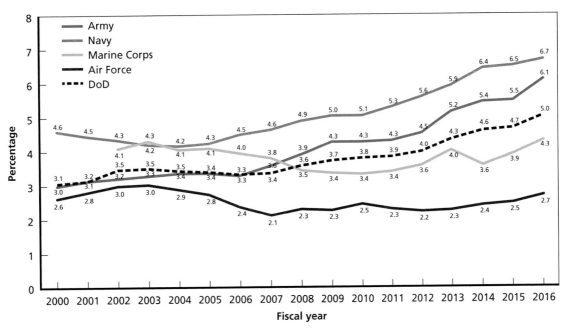

Figure 2.3
Percentage of Active Duty Officers with More Than 30 Years of Service, by Grade

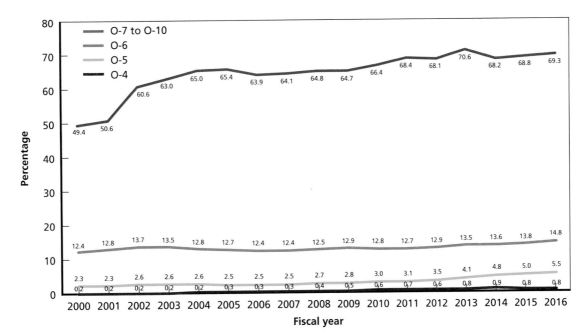

Figure 2.4
Number of O-4 Active Duty Officers with More Than 30 Years of Service

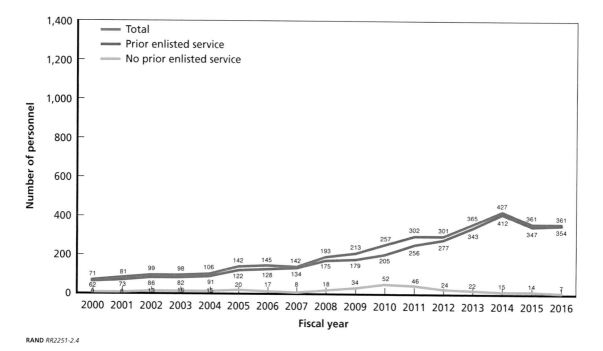

RAND *RR2251-2.4*

Figure 2.5
Number of O-5 Active Duty Officers with More Than 30 Years of Service

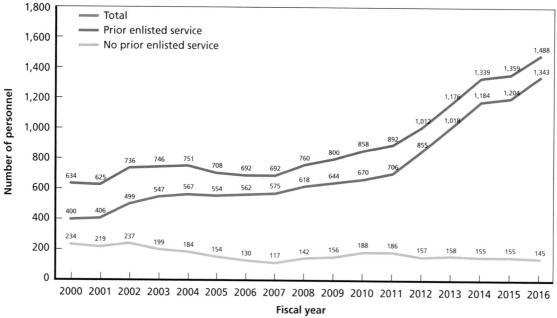

RAND *RR2251-2.5*

Figure 2.6
Number of O-6 Active Duty Officers with More Than 30 Years of Service

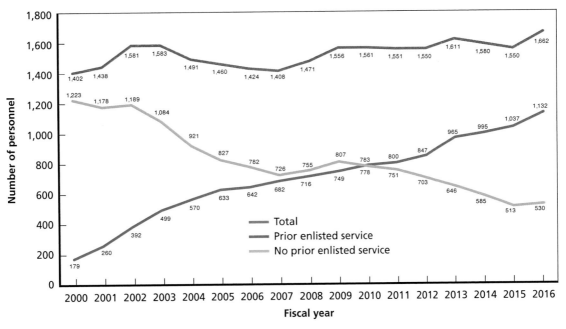

RAND RR2251-2.6

Figure 2.7
Number of O-7 to O-10 Active Duty Officers with More Than 30 Years of Service

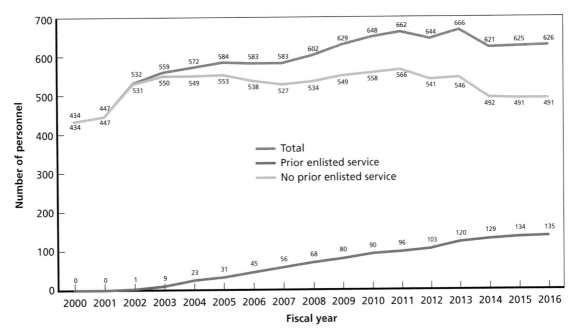

RAND RR2251-2.7

The number of O-5 officers with more than 30 YOS and with prior enlisted service has also grown steadily since 2001, with the rate of increase accelerating between 2011 and 2014. After 2014, the number stabilized in 2015 and then continued to increase in 2016. The number of O-5 officers with more than 30 YOS with no prior enlisted service declined between 2002 and 2007, from 237 to 117 officers; increased to 188 in 2010; and then declined again to 145 in 2016.

The number of O-6 officers with more than 30 YOS and with prior enlisted service also grew between 2000 and 2016, from 179 officers to 1,132. But the number of O-6 officers with no prior enlisted service steadily declined, from 1,223 officers in 2000 to 530 in 2016. Before 2008, most O-6s with more than 30 YOS—87.2 percent in 2000—had no prior enlisted service. After 2008, most had prior enlisted service. By 2016, only 31.9 percent had no prior enlisted service. The interviews we conducted, discussed in Chapter Three, provide some insights into the possible reasons for these trends, including increases in accessions of officers with prior enlisted service to meet past shortages and an increased requirement for the skills and knowledge associated with prior enlisted service.

Our earlier study showed that the number of officers with more than 30 YOS in the grades of O-7 to O-10 increased by 6.5 percent between 2007 and 2014. Figure 2.7 shows the number of such officers in total and whether they have prior enlisted service. Figure 2.8 shows the number of these officers in specific pay grades with no prior enlisted service, and Figure 2.9 shows the number with prior enlisted service.

Most officers in grades O-7 to O-10 with more than 30 YOS have no prior enlisted service. In fact, prior to 2002, none of these officers had prior enlisted service, but that changed in 2002 for O-7s, 2003 for O-8s, 2007 for O-9s, and 2011 for O-10s, as shown at the bottom of

Figure 2.8
Number of O-7 to O-10 Active Duty Officers with No Prior Enlisted Service and More Than 30 Years of Service

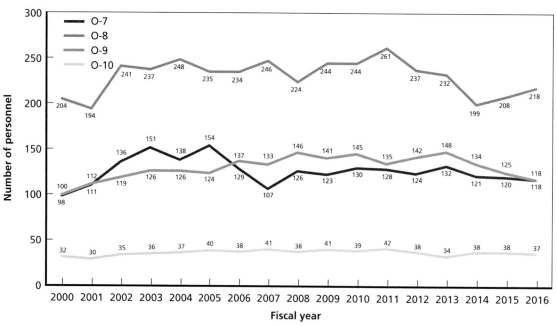

Figure 2.9
Number of O-7 to O-10 Active Duty Officers with Prior Enlisted Service and More Than 30 Years of Service

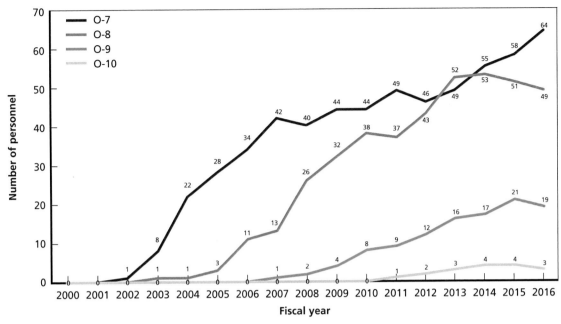

RAND *RR2251-2.9*

Figure 2.9. The staggered introduction of prior enlisted service officers into successively higher grades was no doubt due to the promotion and retention of the lower-grade officers with prior enlisted service. The number of officers in these grades with more than 30 YOS without prior enlisted service increased between 2000 and 2002, remained relatively stable between 2002 and 2013, and then decreased from 546 officers in 2013 to 491 officers in 2016. As shown in Figure 2.8, much of the decline after 2013 shown in Figure 2.7 was due to a decrease in the number of O-7 and O-9 officers with no prior enlisted service.

Tabulations by Service and by Service and Grade

We also consider trends in the number of active duty personnel with more than 30 YOS with prior versus no prior enlisted service by branch of service. In the earlier report, we found that the number of officers with more than 30 YOS began to increase in the Army and Navy even before 2007, but the rate of increase accelerated after 2007. We found little change in the number between 2000 and 2014 in the Marine Corps and little change between 2007 and 2014 in the Air Force, though for the latter service the number dropped significantly prior to 2007.

Figures 2.10 to 2.13 show the number of officers with more than 30 YOS for the Air Force, Army, Marine Corps, and Navy, respectively, by whether they have prior or no prior enlisted service. (We lack data for 2000 and 2001 for the Marine Corps, so the data series begins in 2002). Figures A.1 to A.16 in Appendix A show the number by grade for the Air Force, Army, Marine Corps, and Navy, respectively. We find that the number of officers with more than 30 YOS and with prior enlisted service increased in all four services since 2000. The figures show dramatic increases in the Army and Navy that began before 2007. In the Army, the number

Figure 2.10
Number of Air Force Active Duty Officers with More Than 30 Years of Service

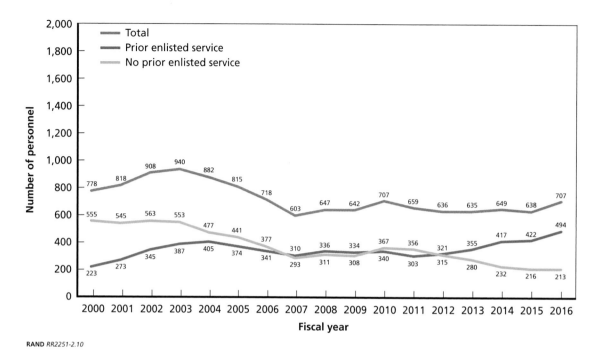

RAND RR2251-2.10

Figure 2.11
Number of Army Active Duty Officers with More Than 30 Years of Service

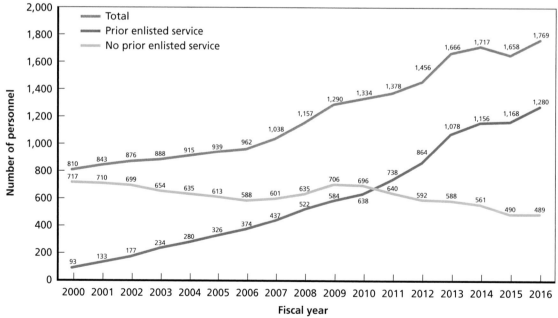

RAND RR2251-2.11

Figure 2.12
Number of Marine Corps Active Duty Officers with More Than 30 Years of Service

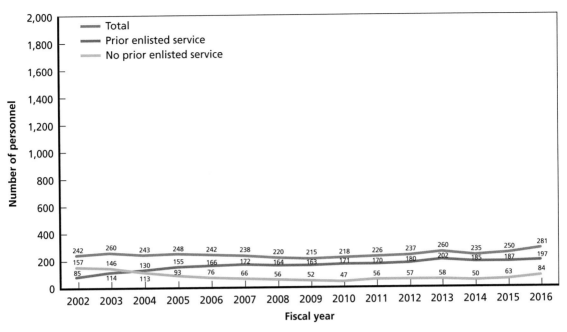

Figure 2.13
Number of Navy Active Duty Officers with More Than 30 Years of Service

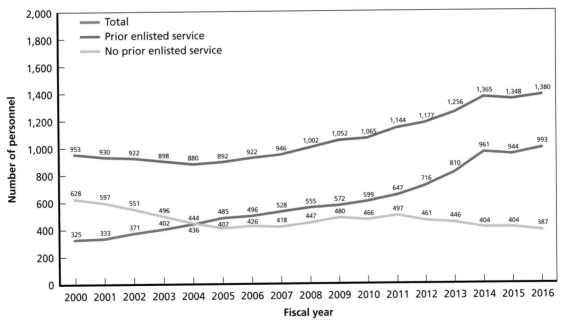

of prior enlisted service officers with more than 30 YOS increased from 93 in 2000 to 437 in 2007 and to 1,280 in 2016, with increases observed for O-4, O-5, O-6, and O-7 to O-10 (Figures A.5–A.8). The number more than tripled in the Navy, from 325 in 2000 to 993 in 2016, with increases especially among O-5 but also among O-4 and O-6 (Figures A.13–A.16). The Air Force also saw an increase between 2000 and 2016, though the increases occurred between 2000 and 2004 and again between 2011 and 2016, with a decrease between 2004 and 2006 followed by a period of relative stability between 2006 and 2011. The number of officers with prior enlisted service also increased in the Marine Corps between 2002 and 2016, though the overall number of officers with more than 30 YOS is quite small relative to the other services.

On the other hand, Figures 2.10 to 2.13 show that the number of officers with more than 30 YOS but with no prior enlisted service decreased in all four services since 2000, with similar decreases by grade, as shown in the figures in Appendix A. Interestingly, the decrease either stopped or reversed for a few years beginning around 2007 in each of the services, before decreasing again. In the Army, the number of officers without prior enlisted service decreased from 717 in 2000 to 588 in 2006, rose to 706 by 2009, but then decreased thereafter, down to 489 by 2016. A similar pattern is seen in the Navy, where the number decreased from 628 in 2000 to 387 in 2016, but there was a short period of increase between 2007 and 2011, followed by a decrease through 2016.

The trends for officers with prior and no prior enlisted service provide an explanation for the trends observed in the total number of officers with more than 30 YOS, found in the earlier study. The accelerated increase in the number of officers with more than 30 YOS after 2007 in the Army and Navy appears to be due the differences in trends before and after 2007 in the number of officers with and without prior enlisted service. The number of officers without prior enlisted service was relative stable, and even increased around 2007, as shown in Figures 2.11 and 2.13. Before 2007, the number of officers with prior enlisted service increased, while the number of officers with no prior enlisted service decreased. The increase in the former group more than offset the decrease in the latter group, for a net increase in the total. But, after 2007, the number of officers without prior enlisted service stopped declining in the Army and Navy, and even increased for a few years, so the rate of increase in the total accelerated. In the case of the Air Force and Marine Corps, the increases in officers with prior enlisted service after 2007 offset the decreases in the number of officers without prior service, so, on net, we find relatively little change in the total after 2007 in these services.

Prior Enlisted Service of Officers with More Than 30 Combined Years of Service

We put some of the tabulations above in context by using DMDC Work Experience File (WEX) data to compute the active duty retention profiles of Army and Navy officers accessed in 1990–1991, and to show the resulting distribution of prior enlisted YOS among officers with more than 30 total active duty YOS in 2016.

The retention profile of officers with prior enlisted service to 20 YOS and beyond is different from that of officers with no prior enlisted experience. Figure 2.14 shows the fraction of officers retained by YOS for a sample of Army active duty officers who were accessed in 1990–1991. The vertical axis on the figure shows the fraction of officers retained, and the horizontal axis shows the combined YOS, counting both prior enlisted service and years of commissioned service. The black stepped line shows the fraction of officers retained by YOS for those officers who enter with no years of prior enlisted service, and it shows the effect of a retirement system with cliff-vesting at 20 YOS, "pulling" officers to 20 YOS and "pushing" officers out thereafter. The red stepped line shows the fraction of officers retained among those who enter with from 1 to 16 years of prior enlisted service, and looks very different, with a higher fraction of officers reaching 20 YOS, and more than 10 percent reaching 30 YOS or more. Figure 2.15 shows similar results for the Navy.

Figure 2.14
Retention of Army Active Duty Officers Accessed
in 1990–1991

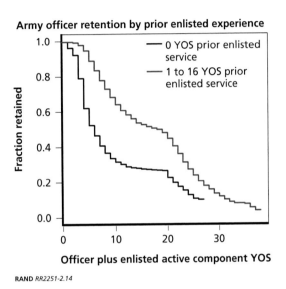

RAND *RR2251-2.14*

Figure 2.15
Retention of Navy Active Duty Officers Accessed
in 1990–1991

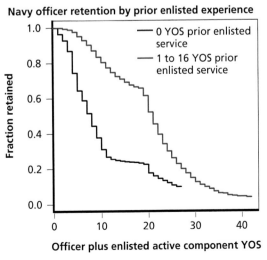

RAND *RR2251-2.15*

Figure 2.16
Distribution of Prior Enlisted Years of Service of Army Active Duty Officers Accessed in 1990–1991 with More Than 30 Years of Service in 2016

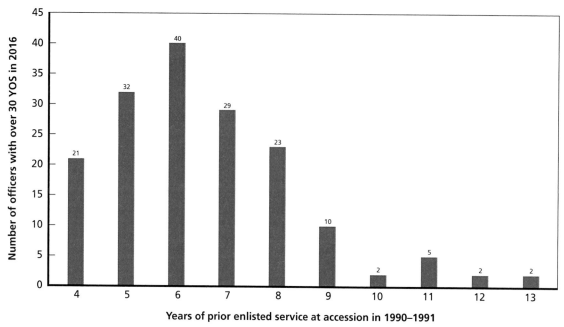

RAND *RR2251-2.16*

Many officers with prior enlisted service and more than 30 combined YOS have served at least one full enlisted term, and some have served two or three. Figure 2.16 shows a 2016 snapshot of the distribution of prior enlisted YOS for Army officers accessed in 1990–1991 with more than 30 combined YOS. The officers shown in the figure were in their 26th or 27th year of commissioned service. The vertical axis shows the number of officers, and the horizontal axis corresponds to the number of prior enlisted YOS; this axis begins at 4 YOS because that would be the minimum necessary to reach 31 or more YOS in 2016 for an officer who began commissioned service in 1990. Each column represents the number of officers for a set number of prior enlisted YOS. For the Army, the mode is at 6 prior enlisted YOS. This contrasts with the Navy, shown in Figure 2.17, where the mode is at 12 prior enlisted YOS. Both services show a wide range for prior enlisted YOS, up to 13 in the Army and 15 in the Navy, with the bulk of the distribution concentrated between 4 and 9 in the Army, and between 9 to 14 in the Navy (with another spike at 5). Thus these officers accessed in 1990–1991 would have appeared as part of the rise in those with prior enlisted service and more than 30 total YOS seen in the previous figures from 2007 to the present.

Summary

This chapter updates the tabulations for officers through 2016 presented in the earlier report and decomposes the tabulations for officers into trends among officers with and without prior enlisted service. The focus of the tabulations is on officers with more than 30 YOS. We find that the number officers with more than 30 YOS continued to increase after 2014, though

Figure 2.17
Distribution of Prior Enlisted Years of Service of Navy Active Duty Officers Accessed in 1990–1991 with More Than 30 Years of Service in 2016

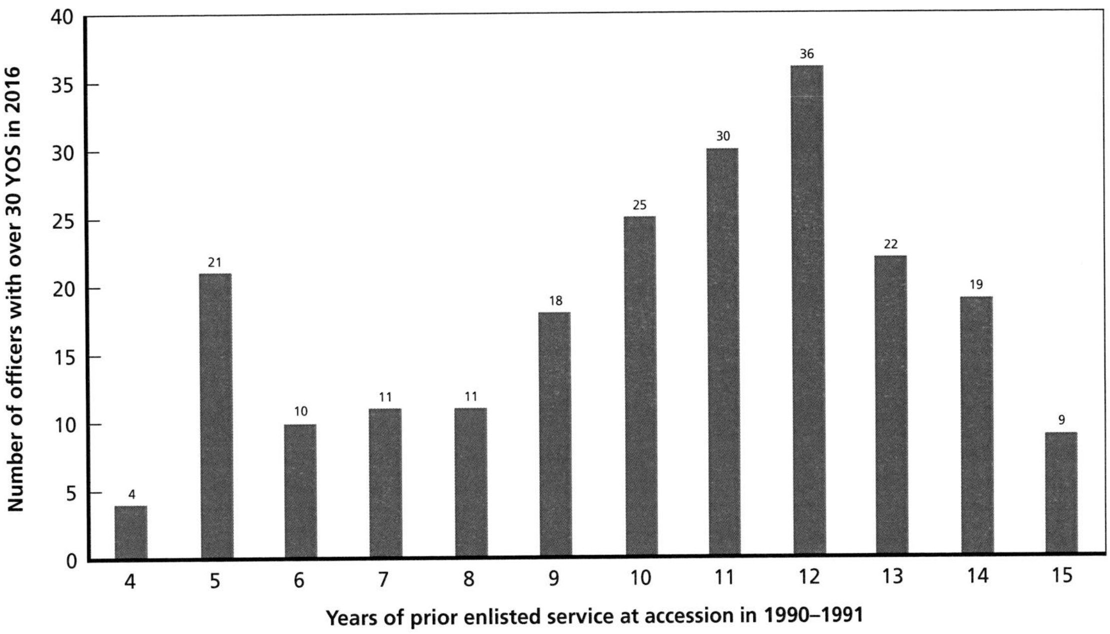

RAND RR2251-2.17

the overall proportion of officers with more than 30 YOS is relatively small across DoD, with only 5 percent of all officers in the grades of O-4 and above as of 2016. The increase in the number of officers with more than 30 YOS is attributable to an increase in officers with prior enlisted service. Our data begin in 2000, and we find that the number with prior enlisted service increased since 2000, while the number without prior enlisted service decreased, though between 2007 and 2013 the number without prior enlisted service was relatively stable. The net effect was an increase in the total number with more than 30 YOS beginning in 2007, thereby explaining the result found in the 2016 study that the number of officers with more than 30 YOS increased markedly after 2007.

We find that nearly all of the officers with more than 30 YOS in grades O-4 and O-5 and the majority in grade O-6 are officers with prior enlisted service. That said, the number of officers in these grades with more than 30 YOS is relatively small, meaning that although most officers with more than 30 YOS are prior enlisted, officers with more than 30 YOS in these grades, especially O-4 and O-5, are relatively rare. By comparison, nearly three-quarters of general and flag officers in the grades of O-7 to O-10 have more than 30 YOS, and the majority of these officers have no prior enlisted service, though the number with prior enlisted service has grown since 2002.

Overall, the number of officers with more than 30 YOS with prior enlisted service has grown since 2000 in all services, while the number without prior service has declined. Among the services, the Army and Navy employ the greatest number officers with prior enlisted service with more than 30 YOS, and the growth in the number with prior enlisted service in these two services has been dramatic. For that reason, our DRM estimation, described in Chapter Four, and the policy simulations we conduct, described in Chapter Five, are for the Army and Navy only.

CHAPTER THREE

Major Themes Emerging from Interviews

Approach

We conducted 14 interviews with senior military officers from each service and the Joint Staff, as well as with civilian DoD employees. The interviewees were in positions of authority relevant to military compensation, retention, and personnel and force management and were familiar with issues related to the 40-year pay table. Some interviewees were experts we interviewed for our 2016 study. The sponsor contacted the interviewees and arranged the interview times. We used a semistructured interview protocol, allowing us to gather the interviewees' perspectives on four broad topics:

1. Why was there an increase in recent years in the number of field grade officers with more than 30 YOS, and what were the roles of such officers with prior enlisted service?
2. Why was there an increase in recent years in the number of NCOs with more than 30 YOS, shown in the 2016 report (Asch et al., 2016)?
3. How would reforming the pay table and retirement formula to allow only members of the highest rank and with the most YOS to earn the highest retirement benefits affect morale, retention, promotion, and force management?
4. Would S&I pay be suitable as a retention tool as an alternative to a retired pay multiplier that increased with YOS beyond 30 years?

A copy of the interview protocol we used is shown in Appendix B.

We sought insight into why the 2007 legislative changes were followed by increases in field grade and senior enlisted personnel with more than 30 YOS, though the intent of the legislation was to increase the retention of general and flag officers. We also wanted the interviewees' thoughts on how morale and force management might be affected by changing the pay table and retired pay formula versus changing S&I pay. The semistructured nature of the interviews allowed us to explore the unique experiences and observations of our interviewees, in both their current and past positions. Few interviewees provided analysis or data to support their observations. Thus, their input was based on their experience, including past analyses they had seen, observation, and speculation. While our experts had diverse experiences and opinions, many of them had common views on the topics above.[1]

---

[1]   We conducted 14 interviews, so we do not have a large enough sample size to make meaningful quantitative inferences about the preponderance of responses. We indicate when only one or two interviewees gave a given response. When we say *few* or *several*, we mean that three to five interviewees gave the response. When we say *many*, we mean the majority gave the response.

Multiple Explanations for the Increase in Experienced Field Grade Officers

A number of interviewees pointed to the increased incentive to serve beyond 30 YOS created by the 2007 legislative changes as a factor contributing to the increase in field grade officers with more than 30 YOS. However, the interviewees also recognized that the increase in the number of field grade officers with more than 30 YOS began before 2007, as shown in the 2016 RAND report and also shown in Chapter Two, and that other, ongoing factors in addition to the 40-year pay table may have contributed to the upward trend. The impacts of these factors could not be disentangled from those of the 40-year pay table legislation. As one interviewee put it, "it's complicated." Interviewees offered a number of explanations for the observed increase, other than the incentive effects of the 2007 compensation changes.

The Great Recession and Changing Requirements for Field Grade Officers

Two interviewees noted that the increase in the number of officers with more than 30 YOS coincided with the Great Recession that began in December 2007. Worsening civilian job opportunities might have induced more officers to remain in the military beyond 30 YOS.

Many interviewees said that the required number of field grade officers has increased over time, implying that the increased number of field grade officers followed increased requirements. The requirements are based on grade, and the quality and expertise of the candidate matters, not YOS per se. One respondent said that Army personnel requirements and promotion processes had been decoupled ten years ago—since then, promotions have been based on the need for and quality of the candidate, rather than the need to promote enough to meet a stated requirement (with candidate quality and experience taking a back seat). This implies that the increase in field grade officers occurred because of the demand for particular individuals' experience, quality, and expertise. A Coast Guard expert said that the Coast Guard does not consider YOS in the decision to fill a billet, but evaluates candidates based on skill levels and grade. Several respondents said that, since service tenure is not a requirement, some positions now filled by officers with more than 30 YOS could be replaced by officers with fewer than 30 YOS. Still, many interviewees stated that the additional years of experience were valuable.

Increased Accessions of Officers with Prior Enlisted Service to Meet Past Shortages

Another factor raised by interviewees as a possible reason for the increase in field grade officers with more than 30 YOS was growth in officer accessions with prior enlisted experience. As an example, they mentioned that in the early 2000s, as well as in the 1990s, when some of the services faced officer shortages, they helped meet their officer accession requirements by commissioning qualified enlisted personnel. Such enlisted personnel can become officers by attending Officer Candidate School/Training or through the Reserve Officers' Training Corps (ROTC) program. Interviewees believed that those with prior enlisted experience comprise a flexible source of officer accessions during times of shortages. After these prior enlisted officers entered the commissioned ranks in the late 1990s and early 2000s, they had an additional financial incentive to stay beyond 30 years, after the 40-year pay table was established in 2007 and additional flexibilities were given by DoD to enable members who would otherwise be mandatorily retired to continue to serve if doing so met the needs of the services.

As discussed in Chapter Two, an officer may serve a limited number of commissioned years before being mandatorily or voluntarily retired under DOPMA. For example, an O-6 who is not on a promotion list to O-7 after 30 years of commissioned service must be retired, except for certain officers in the Navy or Marine Corps who are either limited duty officers (LDOs) or permanent professors at the U.S. Naval Academy. Similarly, an O-5 who is not on a promotion list to O-6 after 28 years of commissioned service must be retired, except for certain officers in the Navy or Marine Corps who are either LDOs or permanent professors at the U.S. Naval Academy. Because enlisted service is not counted as commissioned service, an officer with prior enlisted service could have 30 or more years of total service and fewer than 30 years of commissioned service. Thus, officers with prior enlisted service would not be subject to the DOPMA mandate at 30 YOS for an O-6 and at 28 YOS for an O-5.

Therefore, the increase in field grade officers could be a residual effect from the commissioning of enlisted personnel to meet shortfalls in earlier years, together with the DOPMA rules that are based on commissioned rather than total YOS, according to these experts.

Increased Requirements for the Skills and Knowledge Associated with More Experience, Especially Experience Provided by Prior Enlisted Service

Experts also mentioned an increased requirement for longer careers among those with specialized skills and knowledge. Members with certain expertise were predicted to continue to be in high demand, including pilots and those specializing in cybersecurity and computer science. Such service members typically have lucrative civilian opportunities, so recruiting and retaining them in the military can be difficult. Furthermore, they suggested that the military gets a better return on investment if these personnel stay in military service as long as possible, because their specialized skills are highly valued and require a substantial training investment.

Officers with prior enlisted service were considered by many interviewees to be a key source of personnel with specialized skills. Interviewees stated that having prior enlisted experience made officers better managers of junior enlisted service members because they understood the enlisted experience firsthand. For example, one respondent said that officers with prior enlisted experience are more capable of making time-sensitive decisions as a result of their well-rounded background. Interviewees said that Navy officers with more than 30 YOS are typically in highly technical management fields and tend to be LDOs who get their technical expertise as enlisted members. These officers are the Navy's "middle management" and ensure the Navy's work gets done. Because of the DOPMA rules, many of these officers are able to reach or exceed 30 YOS because they were commissioned with about 10 to 15 years of enlisted experience. The Navy has approximately 5,000 LDOs, with the nuclear career field being the largest.

The additional value of prior enlisted service experience for officers was less clear for the Air Force and the Marine Corps, among experts we interviewed. One expert in Air Force personnel mentioned that there is no productivity difference between those with and without prior enlisted experience. For the Marine Corps, one respondent said that an O-5 is an O-5 regardless of whether the individual has prior enlisted experience. They fill the same billets, meaning that the ones with prior enlisted service do not serve in different types of positions.

Our experts felt that some jobs requiring technical expertise acquired from military service could be replaced with civilians with prior military service, but jobs requiring a deployable capability could not be filled by civilians. For example, in the Navy, LDOs oversee repairs at sea. Furthermore, the interviewees stated that an advantage of military personnel over civilians or contractors is that military personnel can be moved from place to place and can be more flexibly managed, and the military's rank and hierarchical structure provides performance incentives and accountability.

Explanations for the Increase in Experienced Noncommissioned Officers

As with field grade officers, some interviewees felt that the 2007 changes to the military pay table and the other legislative changes contributed to the growth in the number of senior NCOs, and specifically E-9s with more than 30 YOS, because the changes gave individuals an incentive to stay in their current positions longer. Yet, interviewees also mentioned other factors that could have contributed to the growth in experienced E-9s. Several interviewees cited the growing need for senior NCOs to fill positions in the combatant commands, although there is no explicit requirement for them. An Army expert believed the increase in senior NCOs with more than 30 YOS could have been partially due to the creation of Special Volunteer Assistance Brigades, which consist of senior personnel deployed to train foreign military and government forces overseas.

One expert recalled legislative relief in 2007 that relaxed the limit on the number of E-7s and E-8s that could be retained by 0.5 percentage points. This coincided with the conversion to the 40-year pay table and could have contributed to the growth in senior NCOs. For example, these additional E-7s and E-8s could have been promoted to the E-9 level as they progressed in their careers. Another factor mentioned by experts was the growing use of waivers to allow E-9s to stay beyond 30 YOS.

A couple of interviewees thought that changes in the composition of senior NCOs might have generated more of them with greater than 30 YOS. An increase in reservists and recalled retirees serving in the senior enlisted ranks would have increased the number with more than 30 YOS. Both reservists and retirees in the senior enlisted ranks tend to be older with longer service tenure.

Expected Negative Effects of Possible Compensation Changes

We asked interviewees how retention, morale, promotion, and force management might be affected if the 40-year pay table and the multipliers for retirement benefits were changed to allow only members of the highest ranks and with the most YOS to earn the highest retirement benefits. Interviewees expected the effects to be negative, though with some notable exceptions.

Interviewees generally agreed that reducing compensation and/or retirement benefits would have a negative impact on morale and retention. One interviewee speculated that eliminating the expanded retirement benefits might cause a short-term decrease in O-5 retention and a long-term decrease in retention among service members at the beginning of their officer career who are forward-looking and anticipate a lower return on staying in the military. This interviewee predicted that officer strength levels could subside to pre-2007 levels. Several interviewees stated that frequent changes to the pay table and retirement benefit structure would generate uncertainty about future compensation among service members and hurt morale and retention.

With respect to putting different caps on retired pay, depending on grade, the experts believed that capping retired pay for service beyond 30 YOS for those in lower grades would have a negative effect on retention and require offsetting incentives to neutralize the negative effect. One respondent said that this would be "an administrative nightmare" and asked, "How could people plan their careers?" with uncertainty surrounding anticipated retirement benefits. Additionally, everyone interviewed was unanimously against excluding prior enlisted

YOS from the retirement benefit calculation for officers, another potential option that might be considered. They believed it is important to have the same retirement benefit calculation for all officers. Navy experts stated that not crediting officers for prior enlisted service in their retirement calculation would be a disincentive for Navy enlisted members to become LDOs and would hurt officer accessions by sending a signal that their enlisted service is devalued. This type of reform, they argued, would run counter to current efforts to reward enlisted service in the Navy. A Coast Guard expert felt that capping retirement benefits by grade would hurt retention among those in specialized skill positions.

Another disadvantage of capping retired pay by grade that was mentioned is that service members do not have full control over their pay and promotion progression. For example, promotions are affected by whether the force is being downsized or grown, a factor outside the control of individual members. Additionally, service members may stay beyond 30 YOS only if the military allows them to stay, which presumably reflects a readiness requirement. Therefore, limiting the flexibility to provide a retirement incentive to highly valued service members was viewed as having a potentially negative effect on force management.

There were some exceptions to the view that changing the pay table and retirement formula would have negative effects. Within the Marine Corps, only a small number of individuals might be affected by such changes, and retention could be maintained with compensating changes among those in the 30- to 40-YOS range. Another interviewee thought that there would be no meaningful impact on morale or retention because service members with more than 30 YOS would continue to serve anyway, even if they felt short-changed. One expert thought there was a need to cap retired pay to get rid of officers who are "retired on active duty" or in stagnant careers. This interviewee said that managers do not want to give their service members bad personnel evaluations, so these types of officers are allowed to remain on the force. This expert suggested that a better evaluation process was needed to identify who to promote and who should leave.

Special and Incentive Pay Is Not Viewed as a Feasible or Desirable Alternative

Interviewees were not supportive of replacing higher retirement benefits with S&I pay for those with YOS beyond 30. Although doing so was considered more fiscally responsible by some, nearly all interviewees stated that using S&I pay rather than retired pay to induce more retention beyond 30 YOS would have several disadvantages.

First, some interviewees stated that it would be perceived as a cut to military compensation, and the negative public perception of using S&I pay rather than retired pay would have an adverse effect on retention among those with less than 30 YOS. As one interviewee put it, using S&I pay is "a good idea, but a hard journey." Second, several interviewees pointed out that unlike retirement benefits, which are funded by an automatic accrual charge, S&I pay is subject to an appropriation that must be justified annually and competes with other funding priorities, and is therefore not guaranteed. This creates uncertainty for DoD and the services about budgets and for service members about their future pay. By providing future income security, retired pay is potentially more valuable than S&I pay, which is subject to uncertainty, especially as members progress in their careers. Related to the issue of annual budget justification, S&I pay was viewed by some experts as being administratively burdensome.

A third issue raised by one interviewee was the potential for inequities in pay across members in the same grade who perform similar jobs but are in different services. Members could get different S&I pay amounts because each service determines its own S&I pay budget and S&I rates might not be uniform across services, though in some cases they are. The interviewee gave the example of senior enlisted advisers with the same experience and same joint-service job responsibilities who were given different S&I pay amounts depending on their service branch, e.g., an Air Force senior enlisted adviser did not receive the same S&I pay as an Army senior enlisted adviser. One expert pointed out that S&I pay is typically disbursed with a one-year delay, and, as a consequence, S&I pay might not be quick enough to be an effective retention tool.

Finally, one interviewee was not sure whether S&I pay is currently targeted to the right people. For example, Special Operations Forces and other "point of the spear" people get S&I pay, but not the people who support them. The interviewee said, "We need to stop stovepiping by community."

Retired Pay and the 40-Year Pay Table Lend Flexibility to Force Management

Many interviewees talked about the continued need to have flexibility to manage the force and said that the extended pay table and current retired pay formula help to provide such flexibility. They stated that although the increase in the number of personnel, including field grade officers, with more than 30 YOS was unanticipated, it did not mean that it was undesirable. As mentioned, experts felt that the commissioning of qualified enlisted service members through Officer Candidate School or ROTC provided the services with a way to address officer shortages. Furthermore, the interviewees stated that it is desirable to keep some personnel for long careers and that it is therefore important that incentives be in place for them to serve longer.

Long-Term Considerations

Many of the interviewees discussed the importance of considering the long-term horizon when contemplating changes to military compensation. In part, these discussions were meant as a caution against reforming compensation in such a way that could be detrimental to future readiness. An interviewee said that generating short-term savings by cutting retirement benefits, for example, may compromise long-term capability. Another said that the services are currently understaffed and changes that could negatively impact retention should not be considered until staffing level targets have been reached and staffing levels have stabilized. Furthermore, it is advantageous to maintain a surplus of field grade officers to insure against shortages that could be generated by unforeseen deployments and requirements to provide staff to combatant commands.

Interviewees predicted that there will continue to be a growing need for senior personnel with specific types of expertise (e.g., cybersecurity), and the military will have to compete with the civilian sector to retain these individuals. As a result, the current pay and retirement systems must be able to encourage and sustain longer careers in these specialties.

CHAPTER FOUR

Extending the Dynamic Retention Model to Officers with Prior Enlisted Service

The DRM is a model of the service member's decision, made each year, to stay in or leave the active component and, for those who leave, to choose whether to participate in a reserve component and, if participating, whether to continue as a reservist. These decisions are structured as a dynamic program in which the individual seeks to choose the best career path, but the path is subject to uncertainty. The model is formulated in terms of parameters that are estimated with longitudinal data on retention in the active component and participation in the reserve component, and then applied to see how well the estimated model fits observed retention. We use the estimated parameters in policy simulations.

We have described the DRM in earlier documents in which we have estimated a DRM for officers and for enlisted personnel in each service and for selected communities, such as Air Force pilots and military mental health care providers (Asch et al., 2008; Mattock et al., 2016; Hosek et al., 2017). In these earlier efforts, we model the retention behavior of officers with no prior enlisted service, so these earlier models exclude officers with prior enlisted service. Because a high fraction of officers with more than 30 YOS have prior enlisted service and because of the SASC directive to consider policies that would differentially affect officers with prior enlisted service in ways that depend on their pay grade, in this project we extended the DRM capability to analyze the retention behavior of officers with prior enlisted service and to allow for promotion.

Regarding promotion, while an early version of the DRM included grade (Asch et al., 2008), later versions omitted it because adding grade as another "state" that defined a member's career status at a point in time increased the computational burden of the model. Furthermore, the policies that were the focus of our later efforts, namely military retirement reform, were not grade- or promotion-specific, so incorporating grade was not a necessary component of the DRM to assess retirement reform proposals. Because the SASC request involves analysis of grade-specific policies, extending the DRM again to include grade and promotion was necessary. The new model estimates are used to simulate retention and cost effects of retired pay caps, as shown in Chapter Five. This chapter presents an overview of the DRM, describing the extensions to officers with prior enlisted service and to promotion.

In the DRM, a set of parameters underlies the individual officer's retention decisions, and a goal of our analysis is to use individual-level data on officer retention over their careers to estimate the parameters for the Army and for the Navy. We showed in Chapter Two that these two services have the most officers with prior enlisted service and more than 30 YOS. We also discuss the data we use in more detail later in this chapter, but, in short, we use the DMDC's WEX to track individual officer careers from 1990 to 2016. The WEX includes information on prior enlisted service, which allows us to identify which officers have prior enlisted service and the number of years of enlisted service.

Tabulations using the WEX data demonstrate that officers with prior enlisted service have higher retention as commissioned officers than officers without enlisted service. This is seen in Figure 4.1, which shows Kaplan-Meier estimates of cumulative retention by years of commissioned service for Army officers with and without prior enlisted service. By 10 years of commissioned service, 30 percent of officers without prior enlisted service are still in service, compared with 63 percent of officers with prior enlisted service. The higher retention rate of officers with prior enlisted service could be a consequence of higher average taste for the military, as demonstrated by their willingness to make the investment to become an officer and their higher number of total years of active service.

Because of the differences shown in Figure 4.1, we estimate separate models for officers with and without prior enlisted service. As mentioned, our past documentation of the DRM has described the model for officers without prior enlisted service, so our description here focuses on those with prior enlisted service. That said, the fundamental model and key elements of the model are the same.

Model Overview

In the behavioral model underlying the DRM, in each period the individual can choose to continue on active duty, leave the military to hold a job as a civilian, or leave the military to join a reserve component and hold a job as a civilian. The individual bases his or her decision on which alternative has the maximum value. The model assumes that an individual begins his or her military career in an active component.

In past implementations of the DRM, we assumed that officers had no prior enlisted service, so that total years of active service equaled years of commissioned service. In the extension created for this study, officers can begin their commissioned service having had prior enlisted

Figure 4.1
Army Officer Cumulative Retention Rate, by
Whether Officer Has Prior Enlisted Service

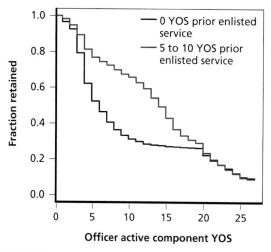

RAND RR2251-4.1

service. Thus, total years of active service no longer necessarily equals total commissioned years, and an officer in the first year of commissioned service may have some number of years of prior enlisted service. The DRM capability could be adapted to model the enlisted member's decision to pursue an officer career, but we do not do that for this study because doing so would significantly expand the scope of the project beyond the time and resources available. We focus on their decision to stay or leave commissioned service, given prior enlisted YOS.

Individuals are assumed to differ in their preferences for serving in the military. Each officer is assumed to have given, unobserved preferences for commissioned active and reserve service, and the preferences do not change. The officer has knowledge of military pay and retirement benefits, as well as civilian compensation. In each period, there are random shocks associated with each of the alternatives, and the shocks affect the value of the alternative. The model explicitly accounts for individual preferences and military and civilian compensation, and in this context, shocks represent current-period conditions that affect the value of being on active duty, being in the selected reserve and being a civilian worker (or *reserve*, for short), or being a civilian worker and not in the reserve (*civilian* for short). Examples of what may contribute to a shock are a good assignment; a dangerous mission; an excellent leader; inadequate training or equipment for the tasks at hand; a strong or weak civilian job market; an opportunity for on-the-job training or promotion; the choice of location; a change in marital status, dependency status, or health status; the prospect of deployment or deployment itself; and a change in school tuition rates. These factors may affect the relative payoff of being in an active component, being in a reserve component, or being a civilian. The individual is assumed to know the distributions that generate the shocks, as well as the shock realizations in the current period but not in future periods.

Depending on the alternative chosen, the individual receives the pay associated with serving in an active component, working as a civilian and serving in a reserve component, or working as a civilian and not serving in a reserve component. In addition, the individual receives the intrinsic monetary equivalent of the preference for serving in an active component or serving in a reserve component. These values are assumed to be relative to that of working as a civilian, which is set at zero.

In considering each alternative, the individual takes into account his or her current state and type. *State* is defined by whether the officer is active, reserve, or civilian and by the individual's active commissioned YOS, reserve commissioned YOS, total years (age), pay grade, prior enlisted YOS, and random shocks. For officers without prior enlisted YOS, this value is set to zero.

Type refers to the level of the individual's preferences for active and reserve service. The individual recognizes that today's choice affects military and civilian compensation in future periods. Although the individual does not know when future military promotions will occur, he or she does know the promotion policy and can form an expectation of military pay in future periods. Further, the individual does not know what the realizations of the random shocks will be in future periods. The expected value of the shock in each state is zero. Depending on the values of the shocks in a future period, any of the alternatives—active, reserve, or civilian—might be the best at the time. Once a future period has been reached and the shocks are realized, the individual can re-optimize (i.e., choose the alternative with the maximum value at that time). The possibility of re-optimizing is a key feature of dynamic programming models that distinguishes them from other dynamic models. In the current period, with future realizations unknown, the best the individual can do is to estimate the expected value of the best choice in the next period, i.e., the expected value of the maximum. Logically, this will also be true in the next period, and the one after it, and so forth, so the model is forward-looking and rationally handles future uncertainty. Moreover, the model presumes that the individual can re-optimize in each future period, depending on the state and shocks in that period. Thus, today's decision takes into account the possibility of future career changes and assumes that future decisions will also be optimizing.

Mathematical Formulation

We denote the value of staying in the active component at time t as

$$V^s(k_t) = V^A(k_t) + \epsilon_t^A,$$

where k_t is defined as

$$k_t = k_t(oy_t, ey_t, ry_t, t, g_t),$$

or the vector of number of active commissioned years oy_t at time t, the number of prior enlisted YOS (ey_t), the number reserve (commissioned) years (ry_t), total years (age), and grade (g_t). $V^A(k_t)$ is the nonstochastic value of the active alternative, and ϵ_t^A is a random shock.

The value of leaving at time t is

$$V^L(k_t) = max[V^R(k_t) + \omega_t^R, V^C(k_t) + \omega_t^C] + \epsilon_t^L,$$

where the member can choose between reserve (R) and civilian (C). *Reserve* means participating in a reserve component and working at a nonmilitary job, and *civilian* means working at a nonmilitary job and not participating in a reserve component. The value of reserve is given by $V^R(k_t) + \omega_t^R$, where k_t is defined above, while value of civilian is given by $V^C(k_t) + \omega_t^C$. We model the reserve/civilian choice as a nest and assume that the stochastic terms follow an extreme value type I distribution, which leads to a nested logit specification.[1] The within-nest shocks to the reserve/civilian choice are given by ω_t^R and ω_t^C, while the nest-level shock is given by ϵ_t^L.

We allow a common shock for the reserve and civilian nest, ϵ_t^L, since an individual in the reserve also holds a civilian job, as well as shock terms specific to the reserve and civilian states, ω_t^R and ω_t^C. The individual is assumed to know the distributions that generate the shocks and the shock realizations in the current period but not in future periods. The distributions are assumed to be constant over time, and the shocks are uncorrelated within and between periods. Once a future year is reached, and the shocks are realized, the individual can re-optimize, i.e., choose the alternative with the maximum value at that time. But in the current period, the future realizations are not known, so the individual assesses the future period by taking the expected value of the maximum, i.e., the expected value of civilian conditional on it being superior to that of reserve times the probability of that occurring, plus the expected value of

---

[1]   See Train (2009) for a discussion of the logit and nested logit specifications.

reserve conditional on it being superior to civilian times the probability of that occurring. For instance, depending on the shocks and the compensation, there is some chance that $V^S(k_t)$ will be greater than $V^L(k_t)$, in which case $V^S(k_t)$ would be the maximum, and vice versa, and the individual makes an assessment of the expected value of the maximum, $Emax(V^S(k_t), V^L(k_t))$.

As noted above, the stochastic terms in the model are assumed to follow an extreme value type I distribution. The extreme value distribution, denoted $EV(a,b)$, has location parameter a and scale parameter b; the mean is $a + b\phi$, and the variance is $\pi^2 b^2/6$, where π is the mathematical constant (~3.14) and ϕ is Euler's gamma (~0.577). As we derived in past studies, this implies

$$\epsilon_t^L \sim EV\left[-\phi\sqrt{\lambda^2 + \tau^2}, \sqrt{\lambda^2 + \tau^2}\right]$$

$$\omega_t^R \sim EV\left[-\phi\lambda, \lambda\right]$$

$$\omega_t^C \sim EV\left[-\phi\lambda, \lambda\right]$$

$$\epsilon_t^L \sim EV\left[-\phi\tau, \tau\right] \ ,$$

where λ is the common scale parameter of the distributions of ω_t^R and ω_t^C, and τ is the scale parameter of the distribution of ϵ_t^L. In the nested structure of the model, leavers face a common shock for the "leave" nest, ϵ_t^L, as well as shocks for the reserve and civilian alternatives within the nest, ω_t^R and ω_t^C, which all together produce a leave shock distributed as extreme value type I, with location parameter $-\phi\sqrt{\lambda^2 + \tau^2}$ and scale parameter $\sqrt{\lambda^2 + \tau^2}$. The logit model requires that the scale parameters of the leave and stay shocks be equal, so we parameterize the model such that the stay scale parameter, which we denote κ, has the same value as the leave scale parameter, i.e., $\kappa = \sqrt{\lambda^2 + \tau^2}$.

The values of the alternatives $V^A(k_t)$, $V^R(k_t)$, and $V^C(k_t)$ depend on the current pay for serving in an active component or working as a civilian, $W^A(k_t)$ or $W^C(k_t)$. An officer with prior enlisted service enters commissioned service at the grade of O-1E, an entry grade that offers higher pay than an O-1, reflecting the prior enlisted service. These officers further progress to O-2E and then to O-3E, and then to O-4, O-5 and so forth. (There are no other grades specific to officers with prior enlisted service beyond O-3E). The officers' active pay is based on total years of active service including both commissioned and noncommissioned YOS, $oy_t + ey_t$.

Our model includes officer promotion. The model assumes that the timing and probability of promotion at each grade are the same for all officers. Promotion to a given grade occurs at a given number of years of commissioned service, but the probability of promotion differs by grade. Also, the probability of promotion is invariant to policy change. Officers that are not promoted face time-in-grade constraints, also called up-or-out constraints, that limit their service career. The combination of not being promoted and up-or-out constraints decreases the value of continuing in the military and operates to decrease retention. Officers that are promoted can look ahead to future promotion gates, and their value of staying is higher than that of officers who are not promoted.

The mathematical expression for the value of the value function $V^A(k_t)$ considers all possible future pathways, recognizing that each pathway depends on each probability of promotion to the next grade and year of commissioned service when promotion can occur. Thus, the DRM views an officer with a particular k_t as reasoning forward to identify the full set of possible future paths of staying or leaving. Then, the officer reasons backward starting from the final stay/leave decision year at the end of each path, called year T.

For each possible k_t, the officer must consider whether to stay or leave. From the perspective of an earlier year t, the officer's current year, there is no reason to commit to a decision at T and in fact it would be short-sighted to do so, because the officer would not be able to base the decision on information that will be revealed when T arrives, i.e., when the shocks in T are realized. Instead, the officer at t develops a decision rule about whether to stay or leave at T, and that rule is to stay if the value of doing so is higher than the value of leaving, and otherwise to leave. The officer can—in the context of the model—compute the expected value of making that optimal decision. Reasoning backward, this expression enters into the expression for the optimal stay/leave decision at $T - 1$, and so on back year by year to t.

At t, the value of continuing in the military for an officer at grade g (now shown as a superscript) is

$$V^A(k_t) = \gamma^A + W_t^{Ag} + \beta\, EMax(V^A(k_{t+1}) + \epsilon_{t+1}^A, V^L(k_{t+1}) + \epsilon_{t+1}^L) + \epsilon_t^A,$$

where γ^A is the individual's taste for active duty, W_t^{Ag}, is active duty pay, β is the personal discount factor, the ϵ terms are random shocks, and the operator $EMax$ finds the expected value of the maximum of the terms $V^A(k_{t+1})$ and $V^L(k_{t+1}) + \epsilon_{t+1}^L$. Each of these terms has a nonrandom term and a random term.

We assume the shocks have an extreme value distribution with a mode of zero and a scale of kappa: $\epsilon \sim EV[0,\kappa]$. With an extreme value shock, the quantity $a + \epsilon$ is distributed as $EV[a,\kappa]$. The mean of this distribution equals the scale factor times Euler's gamma plus the mode: $\phi\kappa + a$, where $\phi \approx 0.577$. If the mode is transformed by subtracting $\phi\kappa$, then $a - \phi\kappa + \epsilon$ is distributed as $EV[a - \phi\kappa,\kappa]$ with a mean of a. Also, if two quantities V^m and V^n have the form $a + \epsilon$ and we subtract $\phi\kappa$ from each, their maximum has an extreme value distribution, namely,

$$Max(V^m, V^n) \sim EV[\kappa\ln(e^{V m/\kappa} + e^{V n/\kappa}) - \phi\kappa, \kappa].$$

The mean, or the expected value, of this distribution of the maximum value of V^m and V^n is

$$\kappa\ln(e^{V m/\kappa} + e^{V n/\kappa}).$$

This result implies that

$$EMax(V^A(k_{t+1}) + \kappa_{t+1}^A, V^L(k_{t+1}) + \kappa_{t+1}^L) = \kappa\ln(e^{V^A(k_{t+1})/\kappa} + e^{V^L(k_{t+1})/\kappa}).$$

To introduce promotion, we replace V^A with its expected value, where p is the probability of promotion:

$$V^A = p_{t+1}^{g+1} V^{A(g+1)} + (1 - p_{t+1}^{g+1}) V^{Ag}.$$

In those years of commissioned service when no promotion occurs, the probability of promotion is zero. In years when promotion occurs, the probability of promotion is assigned a value relevant for the grade.

For simplicity, we assume that civilian pay depends only on commissioned years. If the officer is a reservist, he or she earns the civilian wage plus reserve pay, $W^C(k_t) + W^R(k_t)$. As with active pay, reserve pay depends on total years, including prior enlisted years as well as, of course, reserve years.

The tastes for active and reserve duty, γ^A and γ^R, represent the individual's perceived net advantage of holding an active or reserve position, relative to the civilian state. Other things equal, a higher taste for active or reserve service increases retention. The tastes are assumed to be constant over time but vary across individuals. Also, tastes for active and reserve service are not observed but are assumed to follow a bivariate normal distribution over active component entrants.

The nonstochastic values of the reserve choice and civilian choice can be written as

$$V^R(k_t) = \gamma_r + W^C(k_t) + W^R(k_t) + \beta E[max[V^R(k_{t+1}) + \omega_r, V^C(k_{t+1}) + \omega_c]]$$

$$V^C(k_t) = W^C(k_t) + R(k_t) + \beta E[max[V^R(k_{t+1}) + \omega_r, V^C(k_{t+1}) + \omega_c]],$$

where $R(k_t)$ in the civilian equation is the present value of any active or reserve military retirement benefit for which the individual is eligible. The 2016 NDAA created a new military retirement system, known as the Blended Retirement System. Because our data cover retention decisions of personnel under the legacy retirement system, we use the formula for the legacy system for commissioned officers for the purpose of our analysis given by

$$R(k_t) = 2.5\% \times (oy_t + ey_t) \times W^A(k_t)$$

for the active retirement system where, in this formula, $W^A(k_t)$ is the highest three years of basic pay and is computed based on total active years, $oy_t + ey_t$. For an officer with 30 YOS, the multiplier $2.5\% \times (oy_t + ey_t)$ is 75 percent, while it is 100 percent for an officer with 40 YOS. As discussed in Chapter One, after 2007 the 75 percent cap on the multiplier was lifted, thereby permitting additional YOS beyond 30 to contribute to retired pay.

The model has two switching costs. *Switching cost* refers to a de facto cost reflecting the presence of constraints or barriers affecting the movement from particular states and periods to other states, relative to the movement that would otherwise have been expected from the expressions shown above for the values of staying and of leaving. Switching costs are not actually paid by the individual but, as estimated in the model, are a monetary representation of the constraints or barriers affecting the transition from one state to another at a given time. Further, a switching cost can be either negative or positive. A negative value implies a loss to the individual when changing from the current status to an alternative status, while a positive value implies a gain, or incentive, for the change. The first switching cost is a cost of leaving the active component before the officer's active duty service obligation (ADSO) is completed, assumed to be at YOS 4 for both the Navy and the Army. The estimates, shown later, indicate that the switching cost has a negative value for both services, possibly reflecting the perceived cost of breaching the service contract. The second switching cost is a cost of switching into the reserve from the civilian state. This cost could represent difficulty in finding a reserve position in a desired geographic location or an adverse impact on one's civilian job, e.g., from not being available to work on certain weekends or for two weeks in the summer or being subject to reserve call-up. Its estimated value is negative for both the Army and the Navy.

Estimation Methodology

To estimate DRM, we use the mathematical structure of the model together with assumptions about the distributions of tastes across members and shocks. This allows us to derive expressions for the transition probabilities, given one's state, which are then used to compose an expression for the likelihood of each individual's years of active retention and reserve participation. Importantly, each transition probability is itself a function of the underlying parameters of the DRM. These are the parameters of the taste distribution, the shock distributions, the switching costs, and the discount factor. The estimation routine finds parameter values that maximize the likelihood.

The transition probability is the probability in a given period of choosing a particular alternative, i.e., active, reserve, or civilian, given one's state. Because we assume that the model is first-order Markov, that the shocks have extreme value distributions, and that the shocks are uncorrelated from year to year, we can derive closed form expressions for each transition probability. For example, as Train (2009) shows, the probability of choosing to stay active at time t, given that the member is already in the active component, is given by the logistic form

$$\Pr(V^A > V^L) = \frac{e^{\frac{V^A}{\kappa}}}{e^{\frac{V^A}{\kappa}} + \left[e^{\frac{V^R}{\lambda}} + e^{\frac{V^C}{\lambda}} \right]^{\frac{\lambda}{\kappa}}}.$$

We omit the state vector k_t in each expression for clarity. We can also obtain expressions for the probability of leaving the active component and, having left, the probabilities of entering, or staying in, the reserve component in each subsequent year. To relate the DRM to one-period discrete choice models, we note that in a given period and for a given state and individual taste, the individual's value functions for staying and leaving have the same form as those of a random utility model (McFadden, 1980). Similarly, for those who have left active duty, the choices of whether to enter the reserve or to remain in the reserve are also based on a random utility model. More broadly, the reserve choice is nested in the choice to leave active duty, and the model has a nested logit form. (See Train [2009] for further discussion.) Of course, the DRM differs from a traditional random utility model because the explanatory variables are value functions, not simple variables such as age and education, and the value functions are recursive.

The transition probabilities in different periods are independent and can be multiplied together to obtain the probability of any given individual's career profile of active, reserve, and civilian states that we observe in the data. Multiplying the career profile probabilities together gives an expression for the sample likelihood that we use to estimate the model parameters using maximum likelihood methods. Optimization is done using the Broyden-Fletcher-Goldfarb-Shanno (BFGS) algorithm, a standard hill-climbing method. We compute standard errors of the estimates using numerical differentiation of the likelihood function and taking the square root of the absolute value of the diagonal of the inverse of the Hessian matrix. To judge goodness of fit, we use parameter estimates to simulate retention profiles for synthetic individuals (characterized by tastes drawn from the taste distribution) who are subject to shocks (drawn from the shock distributions), then aggregate the individual profiles to obtain a force-level retention curve and compare it with the retention curve computed from actual data.

We estimate the following model parameters:

- the mean and standard deviation of tastes for active and reserve service relative to civilian opportunities (e.g., μ_a, μ_r, σ_a, and σ_r)
- a common scale parameter of the distributions of ω_t^R and ω_t^C, λ, and a scale parameter of the distribution of ϵ_t^L, or τ
- a switching cost incurred if the individual leaves active duty before completing his or her ADSO
- a switching cost incurred if the individual leaves active and reserve duty before serving a combined total service obligation
- a switching cost incurred if the individual moves from "civilian" to "reserve."

In past DRM analyses, we also estimate a personal discount factor. We fixed the personal discount rates in this study because we found the model fits were better and parameter estimates were more reasonable relative to our expectations based on past research. We set the personal discount factor in this model equal to 0.94, which is the value we have typically estimated for officers in earlier work.

Once we have parameter estimates for a well-fitting model, we can use the logic of the model and the estimated parameters to simulate the active component cumulative probability of retention to each YOS in the steady state for a given policy environment, such as a change to the retired pay cap. By *steady state*, we mean when all members have spent their entire careers under the policy environment being considered. The simulation output includes a graph of the active component retention profile for officers by YOS, where YOS counts prior enlisted service. We can also produce graphs of reserve component participation and provide computations of costs, though we do not do so here. We show model fit by simulating the steady-state retention profile in the current policy environment and comparing it with the retention profile observed in the data.

Data

DMDC's WEX data contain person-specific longitudinal records of active and reserve service. WEX data begin with service members in the active or reserve component on or after September 30, 1990. Our analysis file for the Army and the Navy includes active component officer entrants in 1990 who are followed through 2016, providing 26 years of data for the 1990 cohort. The WEX data provide information on the careers of personnel prior to 1990, thereby allowing us to measure years of prior enlisted service. In constructing the officer samples, we exclude medical personnel and members of the legal and chaplain corps, because their career patterns differ markedly from those of the rest of the officer corps, suggesting that analysis of retention for these personnel needs to be conducted separately. Those in the health professions are the largest of these three groups, making up about 18 percent of all DoD active component officers in FY 2015 (Center for Naval Analyses, undated, Table B-28).

Another key source of data is information on civilian and military pay. For officers, we use the 2007 80th percentile of earnings for full-time male workers with a master's degree in management occupations for civilian pay. The data are from the Census Bureau (undated). Civilian work experience is defined as the sum of active years, reserve years, and civilian years since age 22, but here pay does not vary by other factors, such as years since leaving active duty. Annual military pay for active component officers is represented by regular military compensation (RMC) for FY 2007, equal to the sum of basic pay, basic allowance for subsistence, basic allowance for housing, and the federal tax saved because the allowances are not taxed. Military pay increases are typically across-the-board, with the structure of pay by grade and YOS remaining the same.[2] We therefore do not expect our results to be sensitive to the choice of year. We use RMC tables provided by the Office of the Secretary of Defense (OSD) that show RMC by grade and YOS. Reserve component members are paid differently from active component members, though the same pay tables are used. The method for computing reserve component annual pay is described in Asch, Mattock, and Hosek (2017). Military retirement benefits are related to the basic pay table for officers, and we use the basic pay tables for 2007 for this computation. For officers with prior enlisted service, we use the pay levels for O-1E, O-2E, and O-3E, as appropriate.

We also required data on officer promotion rates and promotion timing to each grade. Officer promotion rates were drawn from those used in Asch and Warner (1994), while promotion timing data were based on computations of average time in service at promotion by grade and service, for FYs 1993 to 2008, from the DMDC. Average time in service at promotion by grade was relatively stable over time, and we used average time in service between 1998 and 2005.

Model Estimates for Officers with No Prior Enlisted Service

Table 4.1 shows the estimated parameters and standard errors for the Army and Navy retention model of officers with no prior enlisted service. To make the numerical optimization easier, we did not estimate most of the parameters directly but instead estimated the logarithm of the absolute value of each parameter, except for the taste correlation, for which we estimated the inverse hyperbolic tangent of the parameter. All of the parameters are statistically significant in the Navy model, and all but the between-nest scale parameter are significant in the Army model. To recover the parameter estimates, we transformed the estimates. Table 4.2 shows the transformed parameter estimates for each service. The estimates are denominated in thousands of 2007 dollars, except for the assumed discount rate and the taste correlation.

We found that mean active taste is negative for the Army and equal to –$24,300. A negative value is consistent with past studies estimating the mean active taste among military officers and suggests that the military must offer relatively high pay to compensate for the require-

---

[2]   An exception was the structural adjustment to the basic pay table in FY 2000 that gave larger increases to midcareer personnel who had reached their pay grades relatively quickly (after fewer years of service). A second exception was the expansion of the basic allowance for housing, which increased in real value between FY 2000 and FY 2005. The costing analysis is in 2017 dollars.

Table 4.1
Parameter Estimates and Standard Errors: Army and Navy Officers with No Prior Enlisted Service

	Army		Navy	
	Estimate	Standard Error	Estimate	Standard Error
Log(Scale Parameter, Nest = τ)	−1.36	33.83	5.20	0.04
Log(Scale Parameter, Alternatives within Nest = λ)	4.69	0.03	3.40	0.06
Log(−1*Mean Active Taste = μ_a)	3.19	0.04	3.00	0.05
Log(−1*Mean Reserve Taste = μ_r)	5.63	0.05	4.01	0.05
Log(SD Active Taste = σ_a)	3.76	0.04	3.87	0.05
Log(SD Reserve Taste = σ_r)	5.26	0.05	3.88	0.06
Atanh(Taste Correlation = ρ)	0.67	0.02	0.94	0.01
Log(−1*Switch Cost: Leave Active <ADSO)	4.81	0.03	5.20	0.04
Log(−1*Switch Cost: Switch from Civilian to Reserve)	6.05	0.03	4.89	0.05
Personal Discount Factor β (Assumed)	0.94	N/A	0.94	N/A
−1*Log Likelihood	24,141		32,139	
N	5,318		6,445	

SOURCE: Parameter estimates from cohorts of personnel entering active duty as officers in 1990–1991.

NOTES: The scale parameter κ governs the shocks to the value functions for staying and for the reserve-versus-civilian nest and equals $\sqrt{\lambda^2 + \tau^2}$. The means and standard deviations of tastes for active and reserve service relative to civilian opportunities are estimated, as are the costs associated with leaving active duty before completing ADSO, and switching from civilian status to participating in the reserve. The personal discount factor was assumed to be 0.94 in these models.

ments of service on active duty relative to not being in the military. For the Navy, the point estimate of mean active taste is negative but smaller in absolute value than for the Army, equal to −$20,060. Both estimates of mean active taste are statistically different from zero.

Mean taste for reserve duty is negative: −$279,980 for Army and −$55,370 for Navy officers. As for the variance in tastes, we found that the standard deviation of active duty taste is larger for the Navy than for the Army, $42,890 for Army officers and $47,770 for Navy officers, while the standard deviation of reserve taste is $191,570 for the Army and $48,660 for the Navy.

The estimated scale parameter for the between-nest shock in the Navy model is much larger than the means and standard deviations of tastes, while the within-nest shock is of the same order of magnitude. These scale parameters provide information on the standard deviation of the common random shock for the reserve/civilian nest, as well as the within civilian/reserve nest shocks. The model nests the reserve and civilian alternatives because most reservists also hold a civilian job; hence, a shock to civilian is also likely to be felt by reserve. The scale parameter for the active and reserve/civilian shock is $\sqrt{\lambda^2 + \tau^2}$, while the within civilian/reserve nest shock is λ. We estimate λ to be $29,960 and τ to be $181,830 for the Navy. These estimates imply that the scale parameter for the total shock, κ, is $184,278. The relative magnitudes of the scale parameters suggest that movement between the active nest and the reserve/civilian nest is largely driven by random shocks rather than by diverse tastes among Navy

Table 4.2
Transformed Parameter Estimates: Army and Navy Officers with
No Prior Enlisted Service

	Army	Navy
Scale Parameter, Nest = τ	0.26	181.83
Scale Parameter, Alternatives within Nest = λ	109.15	29.96
Mean Active Taste = μ_a	−24.30	−20.06
Mean Reserve Taste = μ_r	−279.98	−55.37
SD Active Taste = σ_a	42.89	47.77
SD Reserve Taste = σ_r	191.57	48.66
Taste Correlation = ρ	0.58	0.74
Switch Cost: Leave Active <ADSO	−122.34	−180.42
Switch Cost: Switch from Civilian to Reserve	−425.02	−133.41
Personal Discount Factor β (Assumed)	0.94	0.94

NOTE: Transformed parameters are denominated in thousands of 2007 dollars, with the exception of the taste correlation and personal discount factor. Definitions of variables are described in the text.

members (i.e., taste heterogeneity), while the movement between civilian and reserve statuses is equally driven by diverse tastes and random shocks.

For the Army, we found that τ is small and not statistically significant from zero, so that the scale parameter for the active and reserve/civilian shock is essentially reduced to λ. We estimated a λ of $109,150, approximately four times the estimated mean active taste −$24,300, and about 40 percent of the absolute value of the estimated mean reserve taste −$279,980, implying that tastes, as well as shocks, play a role in explaining shifts into and out of active, reserve, and civilian statuses for the Army.

The switching costs for leaving active duty early, before completing ADSO, are −$122,340 for Army officers and −$180,420 for Navy officers. The cost of switching to a reserve component after being a civilian is −$425,020 for Army officers and −$133,410 for Navy officers. These high costs may reflect the difficulty of finding an available reserve position or an implicit cost to one's civilian career and lifestyle.

Model Estimates for Officers with Prior Enlisted Service

Table 4.3 shows the estimated parameters and standard errors for the Army and Navy retention model of officers with prior enlisted service. Table 4.4 shows the transformed parameter estimates for each service. All of the parameters are statistically significant in the Army model, and all but the mean active taste μ_a are significant in the Navy model.

We found that mean active taste is positive for the Army and equal to $12,570. A positive taste seems plausible given that the individuals who enter as officers with enlisted experience would tend to have higher taste than individuals who had never served in the Army before. For the Navy, the point of estimate of mean active taste is not statistically different from zero,

Table 4.3
Parameter Estimates and Standard Errors: Army and Navy Officers with Prior Enlisted Service

	Army		Navy	
	Estimate	Standard Error	Estimate	Standard Error
Log(Scale Parameter, Nest = τ)	5.60	0.08	4.86	0.20
Log(Scale Parameter, Alternatives within Nest = λ)	4.32	0.26	4.45	0.41
Log(Mean Active Taste = μ_a)	2.53	0.24	−0.37	2.89
Log(−1*Mean Reserve Taste = μ_r)	5.96	0.29	7.88	0.36
Log(SD Active Taste = σ_a)	3.94	0.07	3.80	0.07
Log(SD Reserve Taste = σ_r)	5.40	0.30	7.12	0.47
Atanh(Taste Correlation = ρ)	−0.45	0.01	−0.91	0.29
Log(−1*Switch Cost: Leave Active <ADSO)	5.11	0.11	4.91	0.17
Log(−1*Switch Cost: Switch from Civilian to Reserve)	5.84	0.26	6.37	0.39
Personal Discount Factor β (Assumed)	0.94	N/A	0.94	N/A
−1*Log Likelihood	1,026		778	
N	6,264		2,257	

SOURCE: Parameter estimates from personnel entering active duty as officers 1990–1991.

NOTES: The scale parameter κ governs the shocks to the value functions for staying and for the reserve-versus-civilian nest and equals $\sqrt{\lambda^2 + \tau^2}$. The means and standard deviations of tastes for active and reserve service relative to civilian opportunities are estimated, as are the costs associated with leaving active duty before completing ADSO, and switching from civilian status to participating in the reserve. The personal discount factor was assumed to be 0.94 in these models.

which also seems plausible, as this is still consistent with the idea that individuals entering the Navy as officers with prior experience in the Navy as enlisted would tend to have higher taste than entrants with no Navy experience, whom we estimated had a mean taste of –$20,060. The corresponding standard deviations of active taste are $51,440 and $44,980 for the Army and Navy, respectively.

Mean taste for reserve duty is strongly negative for both the Army and the Navy, at –$388,550 and –$2,646,370, respectively, which makes sense in light of the source data, which show that fewer individuals with enlisted experience who become commissioned officers choose to later participate in the reserve component than their peers with no years of enlisted service, particularly in the Navy. The corresponding standard deviations of reserve taste are $221,530 and $1,246,280 for the Army and Navy, respectively.

The estimated scale parameters for the shocks are larger than the mean and standard deviation of active taste; however, they are smaller than the mean and standard deviation for reserve taste. In the Army, the between-nest scale parameter is $270,240 and the within-nest scale parameter is $75,070, while in the Navy the between-nest scale parameter is $129,300 and the within-nest scale parameter is $85,860. This means that the shock tends to predominate in movement between the active nest and the reserve/civilian nest, while taste tends to predominate for movements within the reserve/civilian nest, i.e., between civilian and reserve component status.

Table 4.4
Transformed Parameter Estimates: Army and Navy Officers with No Prior Enlisted Service

	Army	Navy
Scale Parameter, Nest = τ	270.24	129.30
Scale Parameter, Alternatives within Nest = λ	75.07	85.86
Mean Active Taste = μ_a	12.57	0.69
Mean Reserve Taste = μ_r	−388.55	−2646.37
SD Active Taste = σ_a	51.44	44.98
SD Reserve Taste = σ_r	221.53	1246.28
Taste Correlation = ρ	−0.42	−0.72
Switch Cost: Leave Active <ADSO	−166.25	−135.96
Switch Cost: Switch from Civilian to Reserve	−345.49	−585.78
Personal Discount Factor β (Assumed)	0.94	0.94

NOTE: Transformed parameters are denominated in thousands of 2007 dollars, with the exception of the taste correlation and personal discount factor. Definitions of variables are described in the text.

The switching costs for leaving active duty early, before completing ADSO, are of the same magnitude as those observed for officers with no prior enlisted service, at −$166,250 and −$135,960 for the Army and Navy, respectively. The cost of switching to a reserve component after being a civilian is also similar to that of the officers with no prior enlisted service, at −$345,490 for the Army and −$585,780 for the Navy. Thus, officers with prior enlisted service face implicit costs similar to those faced by the officers with no prior enlisted service.

Model Fit

To assess model fit, we used the parameter estimates to simulate the behavior of synthetic personnel represented by tastes drawn from the active/reserve taste distribution and subject to shocks drawn from a shock distribution with a scale parameter equal to the estimated value. Given active and reserve tastes, current-period shock values, knowledge of the expected pay and promotion environment in the military and the civilian world, and knowledge of the shock scale parameter, each synthetic individual, behaving as a dynamic-program decisionmaker, makes a stay-or-leave decision in each YOS in the active component. This generates a career length of service in the active component. After leaving active service, the individual becomes a civilian and makes a yearly decision regarding reserve participation. If the individual is not in the reserve, the decision is whether to participate; if the individual is in the reserve, the decision is whether to continue to participate. These decisions generate information about reserve participation by year for the years after active component service. We obtained the predicted active component retention profile by adding together these simulated active component retention profiles across a large number of simulated individuals, and we similarly combined individual reserve participation profiles to obtain the predicted reserve participation profile for the population of simulated individuals. The predicted profiles are plotted against the actual profiles to assess goodness of fit. Because both the Army and Navy models for individuals with prior enlisted experience are specific to the number of enlisted YOS, we can generate predicted profiles for any assumed enlisted YOS, as well as a predicted profile that aggregates over the observed distribution of enlisted YOS for individuals who entered as officers in 1990–1991.

Figures 4.2 and 4.3 show the model fit graphs for the active component for the Army and Navy, respectively. The red lines are simulated cumulative retention, and the black lines are retention observed in the data. The figures show the Kaplan-Meier survival curves, and the dotted lines show the confidence interval for the Kaplan-Meier estimates for the observed data.

The horizontal axis counts years since the individual was observed beginning active service. The vertical axis shows the cumulative probability of retention on active duty until that year.

The model fit for the active component is good for the Army and captures the general sweep of Navy retention. In both cases, the simulated retention line lies close to the observed retention line, and reflects the pattern of retention seen in the data with attrition first being high, then slowing after midcareer as vesting in the defined-benefit retirement approaches, and then falling quickly once the vesting point is reached.

Figures 4.4 and 4.5 show composite model fit graphs for active component officers with prior enlisted service, for the Army and Navy, respectively. These graphs show a composite over the empirical distribution of prior enlisted service YOS observed in the data. In both cases, the fit is good, with the simulated retention line lying within or very close to the error bounds of the Kaplan-Meier survival curve.

Figures 4.6 and 4.7 show a few examples for the Army and Navy of model performance in fitting officer active retention histories for particular enlisted YOS.

Figure 4.2
Model Fit Results: Army Officers with No Prior Enlisted Service

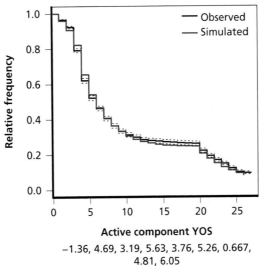

Active component YOS
−1.36, 4.69, 3.19, 5.63, 3.76, 5.26, 0.667, 4.81, 6.05

Figure 4.3
Model Fit Results: Navy Officers with No Prior
Enlisted Service

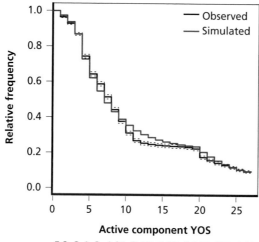

5.2, 3.4, 3, 4.01, 3.87, 3.88, 0.942, 5.2, 4.89

Figure 4.4
Model Fit Results: Army Officers with Prior
Enlisted Service

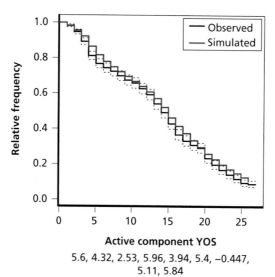

5.6, 4.32, 2.53, 5.96, 3.94, 5.4, −0.447,
5.11, 5.84

Figure 4.5
Model Fit Results: Navy Officers with Prior
Enlisted Service

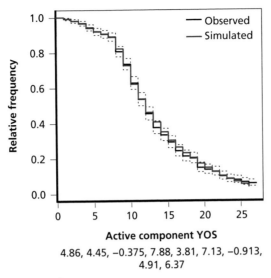

4.86, 4.45, –0.375, 7.88, 3.81, 7.13, –0.913,
4.91, 6.37

Figure 4.6
Model Fit Results: Army Officers with Prior Enlisted Service and 6 or 8 Enlisted Years of Service

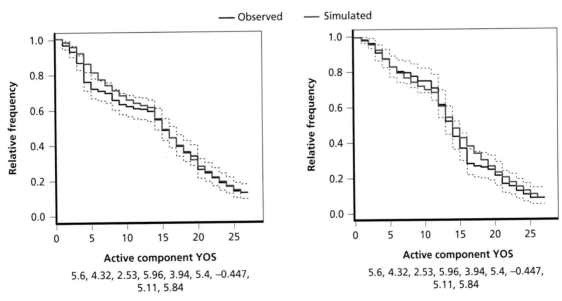

5.6, 4.32, 2.53, 5.96, 3.94, 5.4, –0.447,   5.6, 4.32, 2.53, 5.96, 3.94, 5.4, –0.447,
5.11, 5.84   5.11, 5.84

Figure 4.7
Model Fit Results: Navy Officers with Prior Enlisted Service and 10 or 12 Enlisted Years of Service

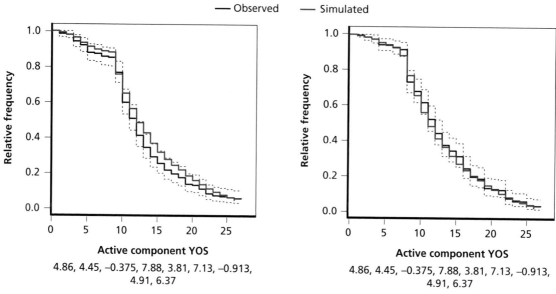

4.86, 4.45, −0.375, 7.88, 3.81, 7.13, −0.913,
4.91, 6.37

4.86, 4.45, −0.375, 7.88, 3.81, 7.13, −0.913,
4.91, 6.37

CHAPTER FIVE

Simulated Retention Effects from the Dynamic Retention Model

The SASC directed DoD to study the effects on retention, cost, morale, promotion, and force management of policies that would cap retired pay based on the highest grade achieved to allow only members in the highest ranks and with the most YOS to earn the highest retirement benefits. It directed the study to consider separate caps for officers and enlisted personnel, to consider measures that would prevent officers with prior enlisted service from using non-commissioned time served to increase their retirement percent eligibility, and to consider the suitability of S&I pay as a retention policy alternative to increasing the retired pay multiplier. Chapter Three discussed these issues, drawing from information from interviews with military manpower policy experts.

In this chapter, we address the retention and cost effects of the SASC directives using the estimated DRM for the Navy and Army, described in Chapter Four. Specifically, we first consider the effects on the retention of Army and Navy officers with prior service, in terms of preventing these officers from counting prior enlisted service in the computation of the retired pay multiplier. Second, we show the retention effects of allowing only members in grades above O-5 to receive credit for service beyond 30 YOS in the computation of the retirement multiplier. That is, for officers O-5 and below, we show the effects of capping the retired pay multiplier at 75 percent (equal to 2.5 percent times 30 YOS) even for those with more than 30 YOS while allowing officers in grades O-6 and above to earn a retired pay multiplier above 75 percent if they have more than 30 YOS. Finally, because capping the retired pay multiplier at 75 percent hurts officer retention, we show the cost effects of restoring retention with a bonus for those who reach 30 YOS.

The analysis in this chapter examines the retention behavior of officers rather than enlisted personnel because the SASC directed that the analysis focus on policies that only allow those in the highest grades and with the most YOS to achieve higher retired pay. As shown in the 2016 RAND report, very few personnel who are E-8 and below serve beyond 30 YOS. That is, virtually all enlisted personnel serving beyond 30 YOS are already in the highest grade, namely E-9. Thus, by outcome, it is already the case the enlisted personnel in the highest grade and with the most YOS earn the highest retired pay multiplier.

The analysis in this chapter also only focuses on Army and Navy officers, because these are the two services with the greatest number of officers with prior enlisted service, as shown in Chapter Two. While officers with prior enlisted service also serve in the Marine Corps, Air Force, and Coast Guard, the numbers are fewer. Given the effort involved in estimating new DRM parameters for prior enlisted officers, discussed in Chapter Four, we focused our analysis on the two services with the most officers with prior enlisted service.

Effects of Not Counting Prior Enlisted Service in the Retired Pay Multiplier Computation

Using the notation from Chapter Four, the current formula for retired pay is

$$R = 2.5\% \times (oy + ey) \times W^A$$

for the active retirement system, where W^A is the highest three years of basic pay ("high-3 pay") and is computed based on sum of commissioned officer years and enlisted years, $oy + ey$. To simplify the notation, the equation drops the time subscript and the k_t argument. We use the DRM to estimate the effect on retention in the steady state of modifying the formula to

$$R = 2.5\% \times (oy) \times W^A,$$

where, under this new formula, retired pay is computed based on commissioned YOS only. Basic pay in the formula is still based on total active YOS, including enlisted years. Thus, if a commissioned officer has 8 years of enlisted service and 22 years of commissioned service, basic pay is computed based on 30 YOS, but the retired pay multiplier is 55 percent (2.5 percent times 22) rather than 75 percent (2.5 percent times 30) under the current system.

We use the DRM to simulate the retention effects for Army and Navy officers with prior enlisted service. (We do not show the results of the effects on officers without prior enlisted service, since the policy obviously would have no effect, given the lack of prior enlisted YOS.) Figures 5.1 and 5.2 show the results for the Army and Navy, respectively. The left panel in each figure shows the cumulative probability of an entrant being retained at each year of active service (including both enlisted and officer YOS), from entry to the officer corps to YOS 40. The figures assume that entry occurs after completing 8 enlisted YOS, though this assumption is only for visual convenience. The computations produced by the DRM allow entry to occur between YOS 4 and 12 for Army officers and between YOS 8 and 14 for Navy officers. Thus, in the figures, at entry (YOS 8), the cumulative probability is 1. Because of the relatively small number of personnel who serve beyond YOS 30, the right panel in each figure zooms in on YOS 30–40. That is, it shows the probability that an entrant reaches YOS 30 to YOS 40, shown on a different scale to highlight differences.

The black line in each graph shows the baseline steady-state retention prior to the change in the retirement formula. The baseline assumes the 40-year pay table, the Executive Schedule Level II cap on basic pay, and the computation of the retired pay multiplier in accordance with the 2014 legislative changes. Importantly, it assumes no cap on the retired pay multiplier, so the retired pay multiplier is 100 percent at 40 YOS (2.5 percent times 40) and assumes that both enlisted and officer YOS are included in the retired pay multiplier calculation.

The red line in each graph shows the simulated effect of not counting enlisted YOS in the retired pay formula for officers with prior enlisted service. It is still the case that the basic pay cap is set to Executive Schedule Level II pay. The simulated retention profiles in both the baseline and under the policy change assume that the Army and Navy made no changes to personnel policy with respect to tightening or loosening their rules on who is permitted to stay. The simulations that produce the profiles also abstract from other changes that occurred, such as changes in the civilian economy or wartime requirements.

Figure 5.1
Simulated Effects on Steady-State Retention of Not Counting Prior Enlisted Service, Army Officers with Prior Enlisted Service

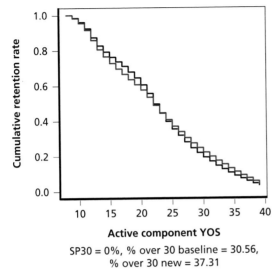

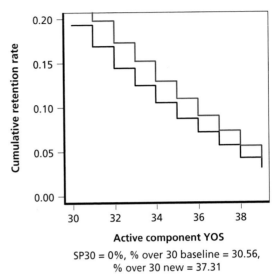

SP30 = 0%, % over 30 baseline = 30.56,
% over 30 new = 37.31

SP30 = 0%, % over 30 baseline = 30.56,
% over 30 new = 37.31

RAND RR2251-5.1

Figure 5.2
Simulated Effects on Steady-State Retention of Not Counting Prior Enlisted Service, Navy Officers with Prior Enlisted Service

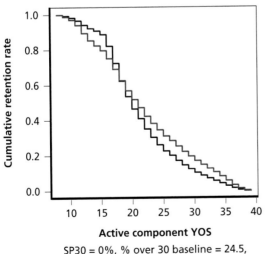

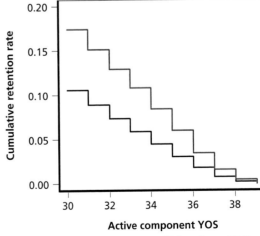

SP30 = 0%, % over 30 baseline = 24.5,
% over 30 new = 37.04

SP30 = 0%, % over 30 baseline = 24.5,
% over 30 new = 37.04

RAND RR2251-5.2

The text at the bottom of each panel shows the number of personnel with more than 30 years of total active service as a percentage of those who have served more than 20 years of active service. Total active years includes both years as an enlistee and years as a commissioned officer. In the baseline, 30.56 percent of Army officers with more than 20 years of total active service have served more than 30 years, while the figure is 24.5 percent for Navy officers with prior enlisted service.

The graphs show that preventing officers from counting enlisted service in the computation of the retired pay multiplier significantly reduces retention before 20 YOS for both the Army and the Navy. That is, while the policy changes directly affected the compensation of those with more than 30 YOS, more-junior members anticipate those changes, affecting their retention decisions before reaching even 20 YOS.

On the other hand, we find that retention after 20 YOS increases for both the Army and the Navy. A member who chooses to serve after 20 YOS forgoes one year of retirement benefits for each year served past year 20, as noted above. *Eligibility* to receive retirement benefits still depends on 20 total YOS, e.g., 8 enlisted YOS and 12 officer YOS, and the retirement benefit is based on high-3 years of basic pay, which in this case are years at officer pay, and total years served, enlisted plus officer. But the retirement benefit is lower because when the officer reaches 20 total YOS the retirement benefit is based on the 12 years of commissioned service as an officer. Therefore, when enlisted service is not counted in the retired pay multiplier, the forgone annuity is lower. However, the additional (incremental) accrued benefit from staying another year remains at 2.5 percent per year. Our results indicate that the former effect dominates the latter effect, meaning that members who are qualified to claim retirement benefits choose to stay longer because they forgo less, on net, by doing so.

We also find that retention after YOS 30 increases among officers with prior enlisted service, as a proportion of those who stay beyond 20 YOS, for both the Army and the Navy. At 30 YOS, the officer has 22 years of commissioned service countable in computing the retirement benefit multiplier and 10 more years before hitting the YOS limit of 40 years. Therefore, there is considerable room for the retirement benefit multiplier to increase. Specifically, we find that the number of personnel with more than 30 YOS as a percentage of those with more than 20 YOS increased from 30.56 percent to 37.31 percent for Army officers and from 24.5 percent to 37.04 percent for Navy officers.

The results indicate that preventing officers with prior enlisted service from using their enlisted service to increase their retired pay multiplier leads to fewer of these personnel staying for a full 20-year career; but those who do stay, stay longer. Our modeling focuses on the retention and not the accession effects of the policy, but the large drop in retention before YOS 20, even for officers in the early part of their officer career, suggests that such a policy would hurt accessions of officers with prior enlisted service as well.

One approach to addressing the lower pre-20-YOS retention would be to offer an S&I pay, say at YOS 16 or YOS 20, assuming it was desirable to return retention to the baseline. While such a policy could close the gap between the red and black lines in the figures for retention before 20 YOS, it would exacerbate the increase in retention after 20 YOS.[1] The increase in post-20-YOS retention might be stemmed by a more stringent application of personnel policies that restrict service beyond 20 YOS and beyond 30 YOS for officers with prior enlisted

---

[1]   We verified this was the case in simulations for the Army and Navy. The results are not shown here.

service. But, the relatively large increase in retention after 20 YOS suggests that such a policy would affect a significant number of members and might be viewed as unfair by the community of officers with prior enlisted service. Furthermore, both current costs and retirement costs increase because retention after 20 YOS increases relative to the baseline.

Effects of Capping the Retired Pay Multiplier by Grade

Using the notation from Chapter Four, the retired pay formula when retired pay is capped for officers who are below grade O-6 with more than 30 YOS is

$$R = Min[(2.5\% \times (oy + ey)), 75\%] \times W^A .$$

Thus, members accrue additional basic pay for service beyond 30 YOS, but the retired pay multiplier does not grow.

For officers in grades O-6 and above, the formula continues to be the baseline formula of

$$R = 2.5\% \times (oy + ey) \times W^A .$$

We use the DRM to simulate the effects on retention and cost for Army and Navy officers, and separately consider officers with and without prior enlisted service. As before, the baseline assumes the 40-year pay table and the Executive Schedule Level II cap on basic pay, no cap on the retired pay multiplier, and the inclusion of both enlisted and officer YOS for computing the multiplier. Under the policy change, retired pay is capped for officers in grades below O-6 with more than 30 YOS. Thus, the policy allows only officers in the highest grades and with the most YOS to get the highest retired pay multiplier.

Figures 5.3 and 5.4 show the results for the Army and Navy, respectively. The graphs show that capping the retired pay multiplier so that it does not increase for years served beyond 30 for those below O-6 reduces steady-state retention for officers with prior enlisted service. Importantly, because members are forward-looking, retention falls not just among personnel with more than 30 YOS but also among those with more than 20 years. That is, while the policy changes directly affect the compensation of those with more than 30 YOS, more-junior members anticipate the effect of those changes on their potential future military earnings. Although they do not know whether they will still be in a grade below O-6 when they reach 30 YOS, they know there is a chance that they will be, thereby affecting their retention decisions before reaching 30 YOS. Specifically, we find that the number of personnel with more than 30 YOS as a percentage of those with more than 20 YOS falls from 30.56 percent to 26.36 percent for Army officers and from 24.5 percent to 19.07 percent for Navy officers.

DOPMA rules require officers in grades O-6 and below and with 30 or more years of commissioned service to separate from service, though waivers are possible. These rules do not apply to officers with prior enlisted service, because these officers do not have 30 years of commissioned service when they reach 30 years of total active service. As shown in Chapter Two, the number of officers with prior enlisted service in grades O-6 and below has increased over time. The simulation results in Figures 5.3 and 5.4 show that capping the retired pay multiplier

Figure 5.3
Simulated Effects on Steady-State Retention of Capping Retired Pay Multiplier by Grade, Army Officers with Prior Enlisted Service

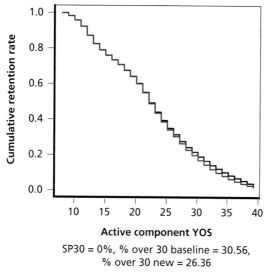
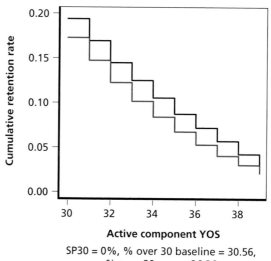

— 40-year Executive II cap, Executive II cap on retired pay, 100 percent multiplier cap
— 40-year Executive II cap, Executive II cap on retired pay, grade-specific multiplier cap

SP30 = 0%, % over 30 baseline = 30.56,
% over 30 new = 26.36

SP30 = 0%, % over 30 baseline = 30.56,
% over 30 new = 26.36

RAND RR2251-5.3

Figure 5.4
Simulated Effects on Steady-State Retention of Capping Retired Pay Multiplier by Grade, Navy Officers with Prior Enlisted Service

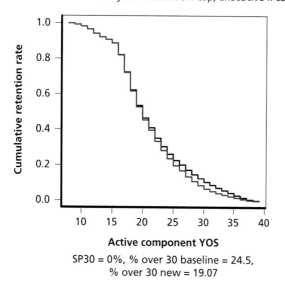
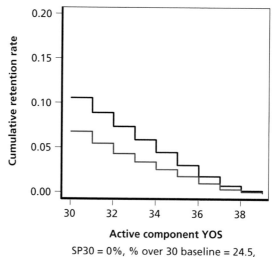

— 40-year Executive II cap, Executive II cap on retired pay, 100 percent multiplier cap
— 40-year Executive II cap, Executive II cap on retired pay, grade-specific multiplier cap

SP30 = 0%, % over 30 baseline = 24.5,
% over 30 new = 19.07

SP30 = 0%, % over 30 baseline = 24.5,
% over 30 new = 19.07

RAND RR2251-5.4

for officers in grades O-5 and below would reduce the number of officers with prior enlisted service serving beyond 30 YOS.

Figures 5.5 and 5.6 show the retention effects of capping the retired pay multiplier for officers *without* prior enlisted service for the Army and Navy, respectively. For these officers, 30 years of active service equals 30 years of commissioned service, so the DOPMA rules for officers in O-6 and below with 30 or more YOS apply to these officers at 30 years of active service. In the baseline, 10.93 percent of Army officers with 20 YOS serve beyond 30 YOS, while 11.84 percent of Navy officers do so. Thus, we find virtually no effect on Navy officers without prior enlisted service and a very small effect for Army officers. It is also relevant to note that officers without prior enlisted service are more likely to retire before 30 YOS as an O-5 and so are less likely to be affected by the cap for O-5s.

The key result is that the effect of capping retired pay for service beyond 30 YOS for officers in grades O-5 and below is smaller for officers without prior enlisted service than for officers with prior service. That is, the red and black lines are nearly identical for officers without prior enlisted service for the Navy, and the red line is just below the black line for the Army.

These results imply that capping the retired pay multiplier so that only officers in the highest grades and with the most YOS receive a higher multiplier would disproportionately affect officers with prior enlisted service. While some officers without prior enlisted service would also be affected, the effect is much more muted.

Whether the drop in retention after YOS 30 poses a problem, whether among officers with or without prior enlisted service, depends on the services' requirements. If the requirements for the most-experienced personnel are lower, say as a result of a military drawdown, then the drop could help achieve a lower requirement. Alternatively, even if requirements are not lower, it is possible that the services could loosen personnel policies that induce mandatory

Figure 5.5
Simulated Effects on Steady-State Retention of Capping Retired Pay Multiplier by Grade, Army Officers without Prior Enlisted Service

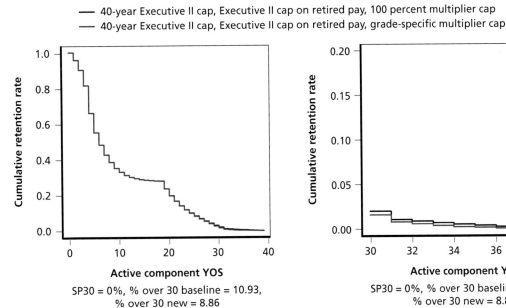
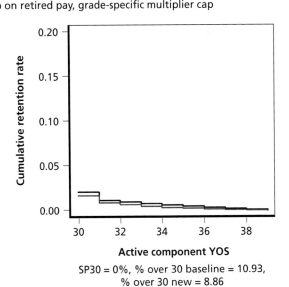

SP30 = 0%, % over 30 baseline = 10.93,
% over 30 new = 8.86

SP30 = 0%, % over 30 baseline = 10.93,
% over 30 new = 8.86

RAND *RR2251-5.5*

Figure 5.6
Simulated Effects on Steady-State Retention of Capping Retired Pay Multiplier by Grade, Navy Officers without Prior Enlisted Service

—— 40-year Executive II cap, Executive II cap on retired pay, 100 percent multiplier cap
—— 40-year Executive II cap, Executive II cap on retired pay, grade-specific multiplier cap

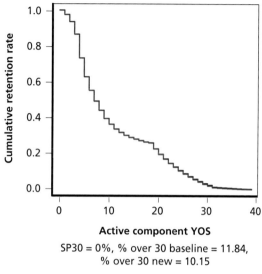 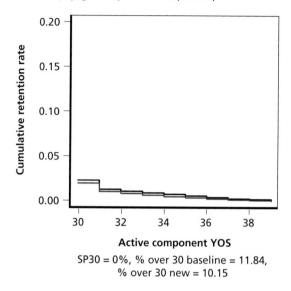

SP30 = 0%, % over 30 baseline = 11.84,
% over 30 new = 10.15

SP30 = 0%, % over 30 baseline = 11.84,
% over 30 new = 10.15

RAND RR2251-5.6

separation, thereby allowing more personnel to stay. That is, the drop in retention might be manageable with other policy tools.

Effects of Offering S&I Pay

Another approach to managing the lower retention associated with the cap on the retired pay multiplier is to offer additional compensation, specifically an S&I pay to sustain retention. We considered a lump sum cash payment paid to members at YOS 30, where the lump sum is computed as a multiplier times annual basic pay at YOS 30. The payment would be made for completion of YOS 29 and paid at the beginning of YOS 30, regardless of whether a member subsequently stayed in service beyond that point.[2] Thus, the S&I pay is like a gate pay: Members receive it for completing a particular milestone—reaching YOS 30. In our analysis, all members reaching this milestone are paid the S&I pay, though in reality, the services might consider targeting the pay to specific members and changing the milestone to a different year, such as YOS 32.

The question of interest is how large the S&I pay multipliers need to be, and how much would need to be budgeted, to restore retention under the retired pay cap policy. That is, what amount of pay is needed to minimize the gap between the black and red lines? We developed an optimization routine in the DRM coding that finds the optimal S&I pay multiplier at YOS 30 to minimize the distance between the red and black lines. Because those with prior service are the majority of officers with more than 30 YOS (Figure 2.1) and the effect of cap-

---

[2]   Our YOS numbering begins with year 0, so in YOS 29 the member is serving in his or her 30th year.

ping the retired pay multiplier is larger for them than for those without prior enlisted service, we calculate the optimized S&I pay multiplier for officers with prior enlisted service. The services could specifically target the S&I pay to officers with prior enlisted service, such as the LDO community in the Navy. Alternatively, the services may choose to offer the S&I pay to all officers who reach 30 YOS, regardless of prior enlisted status. Therefore, we show the effect of offering the S&I pay that is optimal for officers *with* prior enlisted to officers *without* prior service. As we show below, the S&I pay induces more retention than the baseline for officers without prior enlisted service, especially for the Army, implying that these officers would receive a rent, if the bonus were set high enough to sustain the retention of officers with prior enlisted service.

We find that for Army officers with prior enlisted service, the S&I pay multiplier is 1.45, or nearly one and half years of basic pay, in 2017 dollars. The Navy multiplier is a bit higher: 1.62. The retention profiles of Army and Navy officers with prior enlisted service when the retired pay multiplier is capped but when these officers are also paid an S&I pay are shown in Figures 5.7 and 5.8, respectively. The figures show that the retention profiles under the policy change are now virtually identical to the baseline—the red line is nearly on top of the black line. Thus, offering S&I pay sustains steady state retention relative to the baseline for both the Army and the Navy. By implication, the S&I pay almost exactly offsets the decrease in the value of staying at each YOS from 20 to 30 resulting from the retired pay cap policy.

Figures 5.9 and 5.10 show that the multipliers that are optimal for officers with prior enlisted service are too high for officers without prior enlisted service, in the sense that steady-state retention is higher than the baseline. In the case of the Army, offering a multiplier of 1.45 to officers without prior enlisted service—the amount needed to sustain the retention of Army officers with prior enlisted service—would increase retention relative to the baseline before

Figure 5.7
Simulated Effects on Steady-State Retention of Capping Retired Pay Multiplier by Grade and Adding S&I Pay at Year of Service 30, Army Officers with Prior Enlisted Service

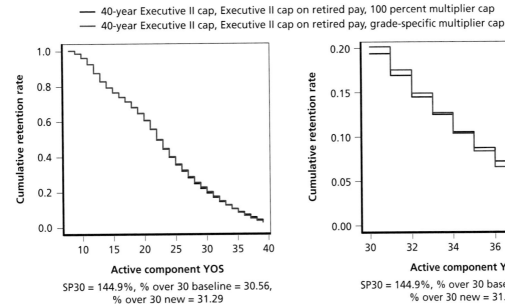

SP30 = 144.9%, % over 30 baseline = 30.56,
% over 30 new = 31.29

SP30 = 144.9%, % over 30 baseline = 30.56,
% over 30 new = 31.29

Figure 5.8
Simulated Effects on Steady-State Retention of Capping Retired Pay Multiplier and Adding S&I Pay at Year of Service 30 by Grade, Navy Officers with Prior Enlisted Service

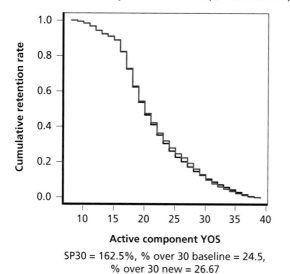
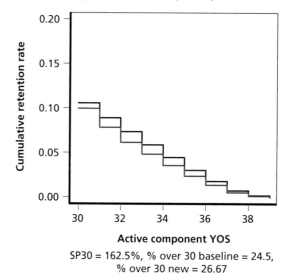

— 40-year Executive II cap, Executive II cap on retired pay, 100 percent multiplier cap
— 40-year Executive II cap, Executive II cap on retired pay, grade-specific multiplier cap

SP30 = 162.5%, % over 30 baseline = 24.5,
% over 30 new = 26.67

SP30 = 162.5%, % over 30 baseline = 24.5,
% over 30 new = 26.67

RAND RR2251-5.8

Figure 5.9
Simulated Effects on Steady-State Retention of Capping Retired Pay Multiplier by Grade and Adding S&I Pay at Year of Service 30, Army Officers without Prior Enlisted Service

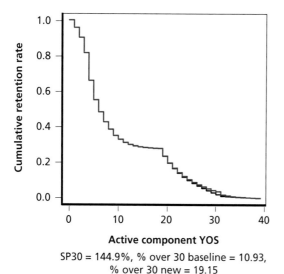
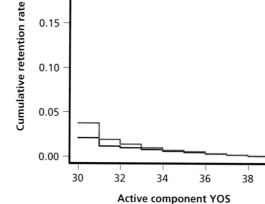

— 40-year Executive II cap, Executive II cap on retired pay, 100 percent multiplier cap
— 40-year Executive II cap, Executive II cap on retired pay, grade-specific multiplier cap

SP30 = 144.9%, % over 30 baseline = 10.93,
% over 30 new = 19.15

SP30 = 144.9%, % over 30 baseline = 10.93,
% over 30 new = 19.15

RAND RR2251-5.9

Figure 5.10
Simulated Effects on Steady-State Retention of Capping Retired Pay Multiplier and Adding S&I Pay at Year of Service 30 by Grade, Navy Officers without Prior Enlisted Service

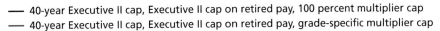

— 40-year Executive II cap, Executive II cap on retired pay, 100 percent multiplier cap
— 40-year Executive II cap, Executive II cap on retired pay, grade-specific multiplier cap

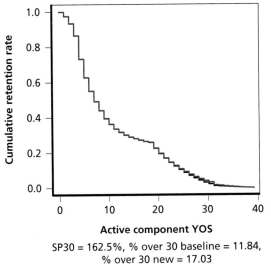

SP30 = 162.5%, % over 30 baseline = 11.84,
% over 30 new = 17.03

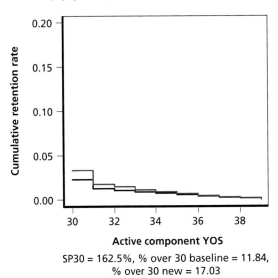

SP30 = 162.5%, % over 30 baseline = 11.84,
% over 30 new = 17.03

RAND *RR2251-5.10*

YOS 30, but especially after 30 YOS. Specifically, we find that the percentage of Army officers with 20 YOS and with no prior enlisted service who serve beyond 30 YOS would increase from 10.93 percent to 19.5 percent. We find a qualitatively similar result for the Navy. The results indicate that officers without prior enlisted service would receive a rent—a payment over and above what is needed to sustain their retention. Therefore, it would be more efficient to target the S&I pay to officers with prior enlisted service, if feasible, should retired pay be capped for officers in grades below O-6.

An alternative approach is to offer an S&I pay to both groups of officers that would preserve their combined retention. Such an approach would make sense if the services were indifferent between officers with and without prior enlisted service when filling a given position. While the experts we interviewed from some of the services, such as the Air Force, said that prior enlisted service status was not important in filling positions, other experts, notably from the Navy and Coast Guard, stated that enlisted experience for officers was highly valued in some occupation areas and communities.[3]

Assuming the S&I pay were targeted to officers with prior enlisted service, we estimated the change in active component cost for these officers. On the one hand, retirement costs are lower because of the retired pay multiplier cap for officers in grades below O-6. That is, officers in these grades who serve more than 30 YOS earn a lower annuity payment. On the other hand, current costs increase because of the S&I pay to officers who reach 30 YOS. For the Army, we estimate that retired pay costs per officer with prior enlisted service decrease

---

[3]   It was not feasible within the project timeline to estimate a DRM that combines the retention behavior of officers with and without prior enlisted service. Consequently, we were unable to estimate the S&I pay that could sustain a combined retention profile.

by 3.4 percent, while current costs per officer increase by 1.8 percent, with a net effect of an increase in active component costs per officer with prior enlisted service of 0.6 percent. For the Navy, retired pay costs per officer with prior enlisted service are estimated to decrease by 2.9 percent, while current costs per officer are estimated to increase by 2.4 percent, with a net effect of an increase in active component costs per Navy officer with prior enlisted service of 0.7 percent. These changes are quite modest, especially given that they only apply to the portion of the officer force with prior enlisted service.

The results indicate that capping the retired pay multiplier with an offsetting increase in S&I pay for officers with prior enlisted service actually increases cost, albeit modestly. The reason is that the S&I pay must be high enough to restore the retention of officers not just with more than 30 YOS but also with fewer than 30 YOS. Furthermore, the S&I pay is paid to those with 30 YOS, and, given that retention falls after 30 YOS, there are more members with exactly 30 YOS than with more than 30 YOS. Thus, the offsetting S&I pay cost exceeds the savings from lower retired pay costs.

CHAPTER SIX

Conclusions

The qualitative and quantitative research in this report supports several conclusions about the feasibility and advisability of capping retired pay. We summarize those conclusions here.

Trends in the Number of Officers with More Than 30 Years of Service

We analyzed retention trends before and after 2007, focusing on officers with more than 30 YOS and decomposing the number of such officers between those who have prior versus no prior enlisted service. Members with more than 30 YOS are a small percentage of the active component officer force—5 percent across DoD in 2016—and the population is relatively small overall. The exceptions are general and flag officers, nearly three-quarters of whom have more than 30 YOS.

The 2016 RAND study (Asch et al., 2016) found that the number of officers with more than 30 YOS increased markedly after 2007, but in percentage terms the greatest increase was not among general and flag officers but among senior enlisted personnel and field grade officers. In the present study, we found that the increase in the number of officers with more than 30 YOS is attributable to an increase in officers with prior enlisted service. We found that nearly all of the field grade officers with more than 30 YOS had prior enlisted service in 2016. The number with prior enlisted service increased since 2000, while the number without prior enlisted service decreased, though between 2007 and 2013 the number without prior enlisted service was relatively stable. The net effect was an increase in the total number with more than 30 YOS beginning in 2007, thereby explaining the result found in the 2016 study. The growth in the number of officers with prior enlisted service was most dramatic in the Army and Navy, and these two services employ the greatest number of officers with prior enlisted service with more than 30 YOS.

DOPMA rules require that field grade officers face mandatory retirement before 30 years of commissioned service if they are not on a promotion list (Title 10 U.S.C., Sections 633 and 634). Because officers with prior enlisted service only become officers after some enlisted service, they will not reach 30 years of commissioned service until they have more than 30 years of total active service. Consequently, these DOPMA rules affect officers with no prior enlisted service at 30 total YOS, but they affect officers with prior enlisted service after 30 total YOS.

The experts we interviewed offered several explanations for the increase in the number of officers with prior enlisted service observed since 2000. Many interviewees said that the requirement for field grade officers has increased over time because of operational needs, and while grade rather than YOS is the key consideration when filling a given billet, several of the service interviewees argued that the requirements for critical expertise, skills, and knowledge has increased over time. These skills are more likely to be found among the most experienced personnel and, in particular, among officers with prior enlisted service, especially technical expertise. The interviewees said that these skills are costly to develop and replace and are sometimes required during deployments, so longer careers are desirable for these personnel. Another reason cited by interviewees was the increase in accessions of officers with prior enlisted service to meet past shortages. In the past, when the services faced officer shortages in the company grades, they accessed more officers with prior enlisted service. Interviewees from the services argued that the ability to access officers who come from the enlisted force provides a flexible source of accessions in times of shortage. Over time, these officers gain commissioned service, but may have only reached the field grades by the time they reach 30 YOS. Some interviewees mentioned that the Great Recession that began in December 2007 worsened civilian opportunities and induced more officers to stay, though this factor would not explain increases in the number of officers after the Great Recession, when the unemployment rate fell from 10.0 percent at the beginning of FY 2010 to 5.0 percent at the beginning of FY 2016 (Bureau of Labor Statistics, undated).

The Effects of Capping Retired Pay

The general view from the experts we interviewed was that capping retired pay so that only those in the highest grades and with the most YOS received the highest retired pay would hurt retention and morale. It could also hurt the accession of officers with prior enlisted service, especially if the implemented cap did not count prior enlisted years in the retired pay multiplier calculation for officers. The interviewees also expressed concern that capping would increase service members' perceived uncertainty about their future retirement benefits and make it more difficult to plan their careers. Furthermore, it would be perceived as unfair because members do not always have full control over their promotion progress—such as during downsizing—and whether their retired pay would be capped.

We extended the DRM capability to officers with prior enlisted service and to allow for the probability of promotions to the next pay grades, and we estimated models for the Army and Navy, the two services with the most field grade officers with more than 30 YOS. We also estimated new models for the Army and Navy officers without prior enlisted service. The models fit the data well, and we used the estimates and the DRM computational coding to develop the capability to simulate the effects of capping retired pay. Specifically, we simulated the retention effects of capping the retired pay multiplier at 30 YOS for officers who reach 30 YOS in grades below O-6, and of preventing the use of prior enlisted service in the computation of the retired pay multiplier.

We found that preventing the use of prior enlisted service in the retired pay multiplier computation significantly changed the experience mix of the force. Retention was lower in the years of service before 20 total YOS, resulting in fewer person-years of service in this range. Retention at 20 YOS was about the same, and retention was higher in the years of service after 20. These results pertain only to officers with prior enlisted service and indicate that the seniority of this group of Army and Navy officers would increase. Offering an S&I pay could prevent lower retention among those with fewer than 20 YOS but would further increase retention after 20 YOS.

We also simulated the effect of capping the retired pay multiplier for officers in grades O-5 and below. We found that retention of officers with prior enlisted service would decline, not only for those with more than 30 YOS but also among those with between 20 and 30 YOS. Officers are forward-looking. Although they do not know whether the cap at 30 YOS will apply to them, there is a chance that it will do so, thereby negatively affecting their current retention decision. On the other hand, the effects of capping retired pay for officers with no prior enlisted service were much more muted. DOPMA rules affect officers with no prior enlisted service at 30 YOS, but they affect officers with prior enlisted service after 30 YOS.

The Feasibility and Advisability of Capping Retired Pay

One approach to managing the lower retention of officers associated with capping the retired pay multiplier is to offer an S&I pay to sustain retention. We find that an S&I pay could sustain the retention of officers with prior enlisted service with a relatively small increase in net cost to the services, if the approach taken for capping retired pay was to cap the multiplier for officers in grades O-5 and below. But S&I pay would actually increase officer retention after 20 YOS if instead the approach taken was to cap the multiplier so that it did not count prior enlisted service. Furthermore, while interviewees broadly agreed that S&I pay could be a useful tool, the general view was that S&I pay was not a feasible or desirable alternative to retired pay. They thought it would be perceived by service members as a cut in military compensation, especially on top of recent military retirement reform, and negative perceptions could hurt retention, thereby increasing the need for S&I pay—and the cost—of sustaining retention. Furthermore, retired pay was considered potentially more valuable than S&I pay, which is subject to uncertainty, especially as members progress in their careers.

Beyond the feasibility of using S&I pay as an alternative retention tool, none of the interviewees argued for limiting the service of field grade officers beyond 30 YOS, especially given the relatively small proportion of field grade officers with more than 30 YOS. As one interviewee put it, while the increase in the number of field grade officers with more than 30 YOS was unexpected, that did not mean it was undesirable. They argued that DoD needs the flexibility to keep some personnel for long careers, and so it is important that the incentives are in place for them to want to serve longer. The view was that there is a continued need to have flexibility to manage the force and retain people with critical skills, and that the extended pay table and current retired pay cap provide such flexibility. Thus, many of the interviewees said it was important to consider the long-term horizon when contemplating changes to military compensation and cautioned against reforming compensation in such a way that could be detrimental to future readiness. Put differently, nearly all stated that capping retired pay was not advisable.

Tabulations of Personnel Strength, by Service and Grade

Figure A.1
Number of O-4 Active Duty Air Force Officers with More Than 30 Years of Service

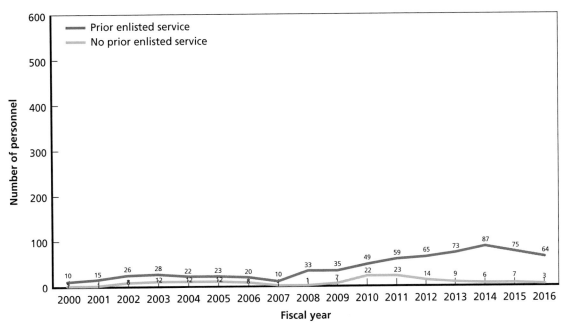

RAND *RR2251-A.1*

Figure A.2
Number of O-5 Active Duty Air Force Officers with More Than 30 Years of Service

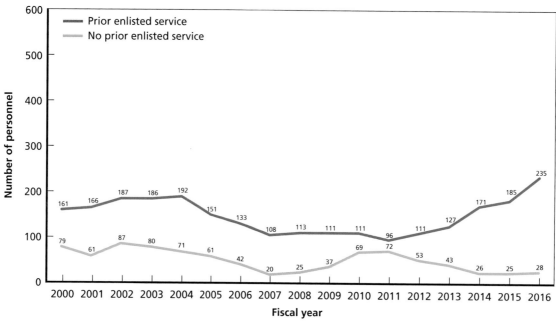

Figure A.3
Number of O-6 Active Duty Air Force Officers with More Than 30 Years of Service

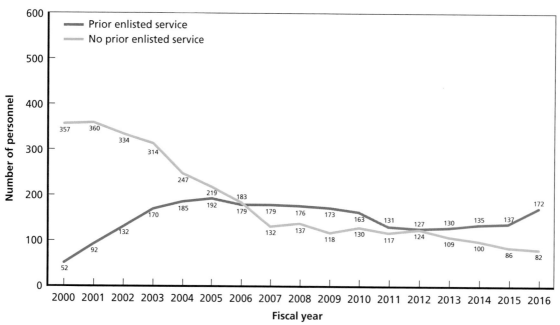

Figure A.4
Number of O-7 to O-10 Active Duty Air Force Officers with More Than 30 Years of Service

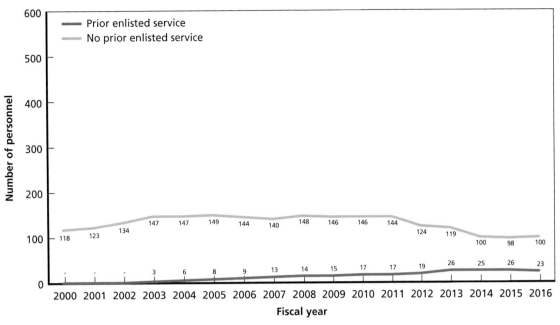

Figure A.5
Number of O-4 Active Duty Army Officers with More Than 30 Years of Service

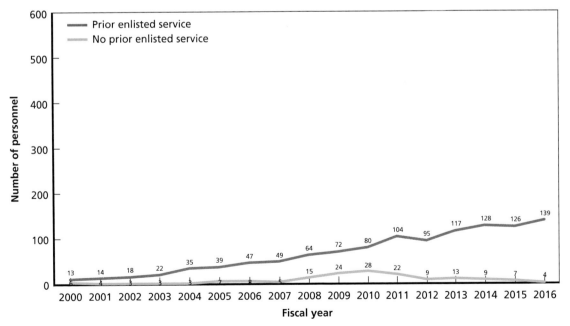

Figure A.6
Number of O-5 Active Duty Army Officers with More Than 30 Years of Service

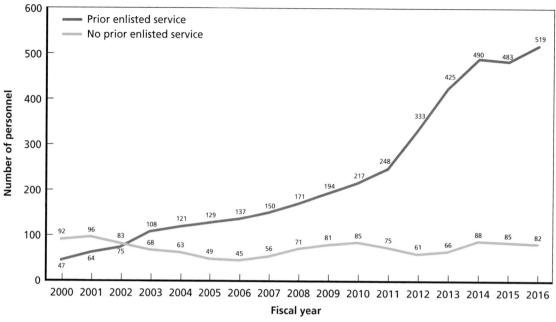

RAND RR2251-A.6

Figure A.7
Number of O-6 Active Duty Army Officers with More Than 30 Years of Service

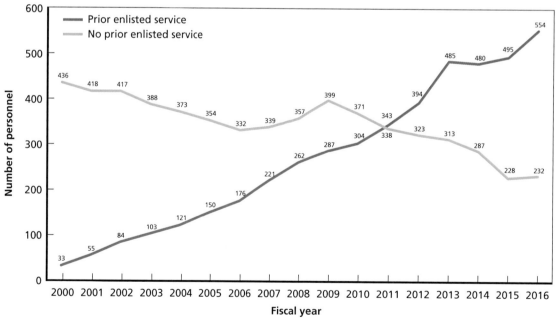

RAND RR2251-A.7

Figure A.8
Number of O-7 to O-10 Active Duty Army Officers with More Than 30 Years of Service

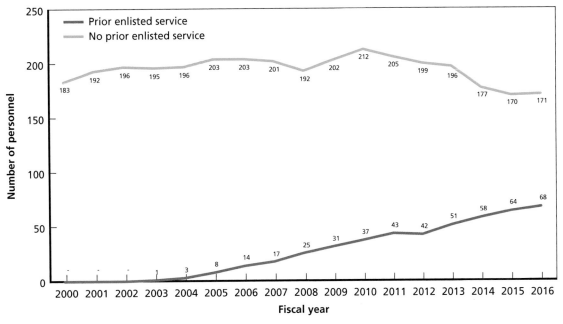

RAND *RR2251-A.8*

Figure A.9
Number of O-4 Active Duty Marine Corps Officers with More Than 30 Years of Service

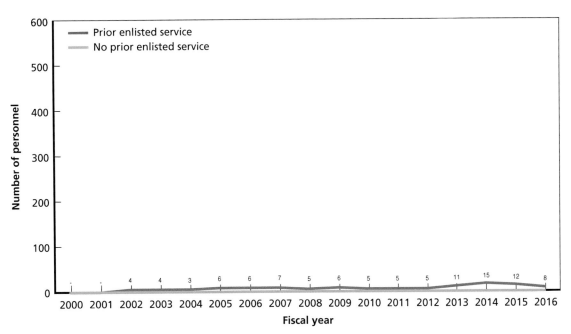

RAND *RR2251-A.9*

Figure A.10
Number of O-5 Active Duty Marine Corps Officers with More Than 30 Years of Service

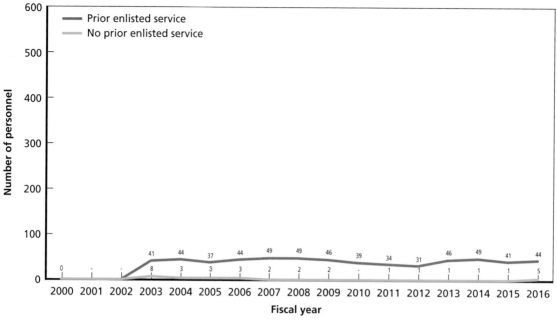

RAND *RR2251-A.10*

Figure A.11
Number of O-6 Active Duty Marine Corps Officers with More Than 30 Years of Service

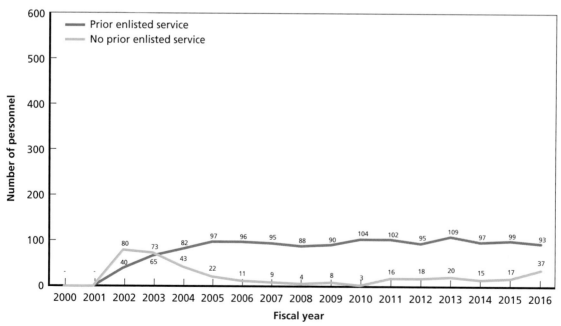

RAND *RR2251-A.11*

Figure A.12
Number of O-7 to O-10 Active Duty Marine Corps Officers with More Than 30 Years of Service

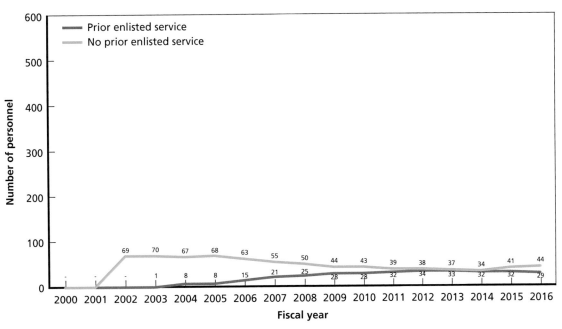

RAND *RR2251-A.12*

Figure A.13
Number of O-4 Active Duty Navy Officers with More Than 30 Years of Service

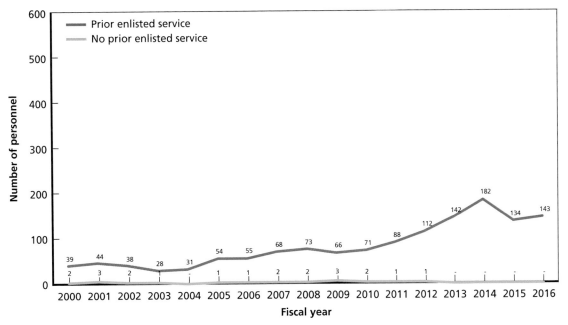

RAND *RR2251-A.13*

Figure A.14
Number of O-5 Active Duty Navy Officers with More Than 30 Years of Service

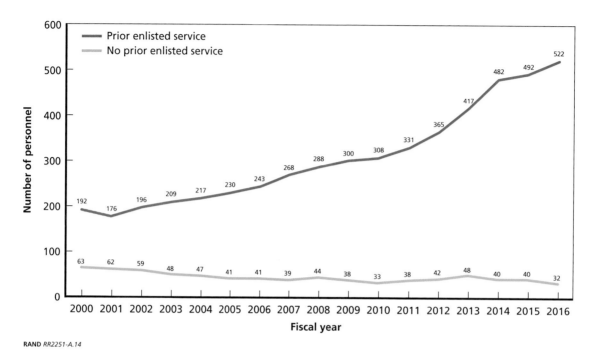

RAND RR2251-A.14

Figure A.15
Number of O-6 Active Duty Navy Officers with More Than 30 Years of Service

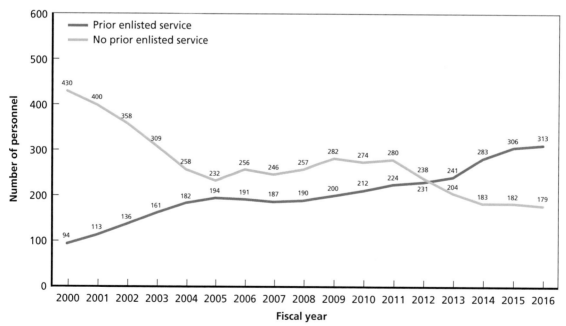

RAND RR2251-A.15

Figure A.16
Number of O-7 to O-10 Active Duty Navy Officers with More Than 30 Years of Service

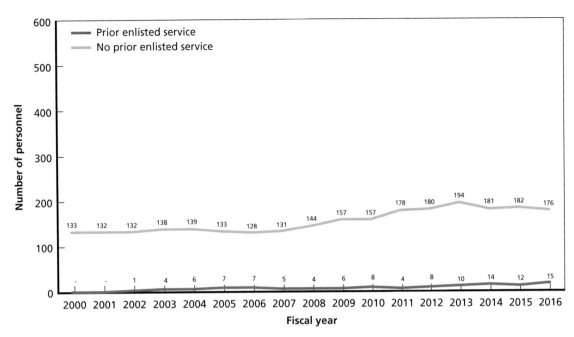

Interview Protocol

SUPPORT FOR DEFENSE REVIEW OF THE MILITARY PAY TABLES

Semi-Structured Interview Protocol for Officers and Officials with Oversight in
Military Personnel Management

Interviewer Introduction and Consent: Thank you for taking the time to speak with
us today. Our conversation will probably take around 30–45 minutes, and we have some
information we would like to give you before we get started.

The Senate Armed Services Committee (SASC) Report that accompanied the FY2017
NDAA directed the Secretary of Defense to conduct a review of the advisability and feasibility of reforming the 40-year pay table and the retirement benefit formula to cap retired pay
based on the highest grade achieved, regardless of years of service, so that only members of
the highest rank and higher years of service earn the highest retirement benefits. OSD(P&R/
MPP) asked RAND to assist with the review, and the purpose of this interview is to gather
information for our study.

The review the SASC requested is a follow on to an earlier study that the SASC
requested and that DoD provided last year to ascertain whether there was utility to retaining
the pay table out to 40 years of service. In the earlier report, DoD stated it opposed reverting
back to the 30-year pay table because the 40-year pay table accomplishes the Department of
Defense's objectives of fielding an experienced and ready force capable of retaining its most
senior members. The analysis in the report showed an increase in the number of military personnel with more than 30 years of service since 2007 when the 40-year pay table came into
effect; the most significant percentage increases were among field grade officers, though the
number of general and flag officers and senior non-commissioned officers also increased. The
report showed that the Department could save $1.2 billion per year by reverting back to the
30-year pay table and by using special and incentive pays to sustain retention after 30 years
of service.

The SASC has directed that DoD assess reforms to the 40-year pay table and to the
cap on retirement benefits to limit retirement benefits for members of lower ranks and
higher years of service. The assessment should consider cost-savings measures that would
allow members to retire with 20 or more years of total service but prevent officers with prior
enlisted service to use non-commissioned time served to increase their retirement percent
eligibility. It also directed that DoD consider the suitability of special pay or bonuses as a
retention tool as an alternative to increasing the retired pay multiplier, to compensate spe-

cific occupational specialties such as chaplains and limited duty officers, specialties that have limited promotion rates but great longevity benefits. In addition to cost savings, the SASC directed that the study assess the impact to morale and retention, and any effects on promotion rates and benefits to force management.

We are interviewing you as a military personnel management expert to learn about the performance of the 40-year pay table vice a 30-year table, the requirements for military personnel with more than 30 years of service and to get your professional assessment of the morale, retention, and cost effects of reforms that that would allow only members of the highest rank and higher years of service to earn the highest retirement benefits.

Your participation is completely voluntary, and you can choose to not answer any specific questions or end the interview at any time. The amount of information you share is completely up to you. Additionally, you will not be identified directly or indirectly in the products of this research, e.g., in briefings or reports. This means we will not use your name or cite your office, although we will describe you and other interviewees collectively as individuals who have, or have had, high-level responsibility for military personnel management and compensation in the military services or the Office of the Secretary of Defense. If you have concerns about your participation in this interview, you may contact the project leaders (Beth Asch, James Hosek, or Michael Mattock), or RAND's Human Subjects Protection Committee [provide copy of the above consent paragraph and contact information]. With that in mind, do you wish to continue with the interview? [Wait for affirmation].

Just so you know, this is a semi-structured interview, meaning that we have some specific questions to get the ball rolling and we'd like you to share any information you think is pertinent to them. The questions come from topics SASC would like the study to address.

1. *To what would you attribute the increase in the number of field grade officers with more than 30 years of service in recent years?*
    a. *Have requirements changed in such a way that the number of field grade officers with more than 30 years of service has increased?*
    b. *Has the availability of officers changed in such a way that the number of field grade officers with more than 30 years of service has increased?*
    c. *Could field grade officers with fewer than 30 years of service perform the same duties?*

2. *To what would you attribute the increase in the number of non-commissioned officers with more than 30 years of service in recent years?*

3. *What are the roles of officers with prior enlisted service who serve with more than 30 years of service? Have those roles changed over time?*

4. *Finally, SASC wondered what would happen if the pay table and retirement formula cap were reformed to allow only members of the highest rank and highest years of service to earn the highest retirement benefits. How do you think this would affect*
    a. *Morale and retention?*
    b. *Promotion?*
    c. *Force management?*

Abbreviations

ADSO	active duty service obligation
DMDC	Defense Manpower Data Center
DoD	U.S. Department of Defense
DOPMA	Defense Officer Personnel Management Act
DRM	Dynamic Retention Model
FY	fiscal year
LDO	limited duty officer
NCO	noncommissioned officer
NDAA	National Defense Authorization Act
OSD	Office of the Secretary of Defense
ROTC	Reserve Officers' Training Corps
SASC	Senate Armed Services Committee
S&I	special and incentive
WEX	Work Experience File
YOS	year(s) of service

References

Army Regulation 601-10, *Management and Recall to Active Duty of Retired Soldiers of the Army in Support of Mobilization and Peacetime Operations*, Washington, D.C.: Department of the Army, March, 13, 2009. As of December 1, 2017:
http://dopma-ropma.rand.org/pdf/AR601-10.pdf

Asch, Beth J., James Hosek, Jennifer Kavanagh, and Michael G. Mattock, *Retention, Incentives, and DoD Experience Under the 40-Year Military Pay Table*, Santa Monica, Calif.: RAND Corporation, RR-1209-OSD, 2016. As of December 1, 2017:
http://www.rand.org/pubs/research_reports/RR1209.html

Asch, Beth J., James Hosek, Michael G. Mattock, and Christina Panis, *Assessing Compensation Reform: Research in Support of the 10th Quadrennial Review of Military Compensation*, Santa Monica, Calif.: RAND Corporation, MG-764-OSD, 2008. As of December 1, 2017:
http://www.rand.org/pubs/monographs/MG764.html

Asch, Beth J., Michael G. Mattock, and James Hosek, *The Blended Retirement System: Retention Effects and Continuation Pay Cost Estimates for the Armed Services*, Santa Monica, Calif.: RAND Corporation, RR-1887-OSD/USCG, 2017. As of December 5, 2017:
http://www.rand.org/pubs/research_reports/RR1887.html

Asch, Beth J., and John T. Warner, *A Policy Analysis of Alternative Military Retirement Systems*, Santa Monica, Calif.: RAND Corporation, MR-465-OSD, 1994. As of December 1, 2017:
http://www.rand.org/pubs/monograph_reports/MR465.html

Bureau of Labor Statistics, "Labor Force Statistics from the Current Population Survey: Unemployment Rate," undated. As of December 1, 2017:
https://data.bls.gov/timeseries/LNS14000000

Center for Naval Analyses, *Population Representation in the Military Services: Fiscal Year 2015*, Alexandria, Va., undated. As of December 1, 2017:
https://www.cna.org/research/pop-rep

Department of Defense Instruction 1320.08, *Continuation of Commissioned Officers on Active Duty and on the Reserve Active Status List*, Washington, D.C.: U.S. Department of Defense, March 14, 2007. As of December 1, 2017:
http://dopma-ropma.rand.org/pdf/DODI-1320-08.pdf

DoD—*See* U.S. Department of Defense.

Hosek, James, Shanthi Nataraj, Michael G. Mattock, and Beth J. Asch, *The Role of Special and Incentive Pays in Retaining Military Health Care Providers*, Santa Monica, Calif.: RAND Corporation, RR-1425-OSD, 2017. As of December 1, 2017:
http://www.rand.org/pubs/research_reports/RR1425.html

Kapp, Lawrence, *Military Officer Personnel Management: Key Concepts and Statutory Provisions*, Washington, D.C.: Congressional Research Service, 7-5700, R44496, May 10, 2016. As of December 1, 2017:
https://fas.org/sgp/crs/natsec/R44496.pdf

Mattock, Michael G., James Hosek, Beth J. Asch, and Rita Karam, *Retaining U.S. Air Force Pilots When the Civilian Demand for Pilots Is Growing*, Santa Monica, Calif.: RAND Corporation, RR-1455-AF, 2016. As of December 1, 2017:
http://www.rand.org/pubs/research_reports/RR1455.html

McFadden, Daniel, "Econometric Models for Probabilistic Choice Among Products," *Journal of Business*, Vol. 53, No. 3, Part 2, pp. S13–S29, 1980.

Public Law 96-513, Defense Officer Personnel Management Act, December 12, 1980.

Public Law 109-364, John Warner National Defense Authorization Act for Fiscal Year 2007, October 17, 2006.

Public Law 113-291, Carl Levin and Howard P. "Buck" McKeon National Defense Authorization Act for Fiscal Year 2015, December 19, 2014.

Public Law 114-92, National Defense Authorization Act for Fiscal Year 2016, November 25, 2015.

Public Law 114-328, National Defense Authorization Act for Fiscal Year 2017, December 23, 2016.

Train, Kenneth, *Discrete Choice Methods with Simulation*, 2nd ed., Cambridge, Mass.: Cambridge University Press, 2009. As of December 1, 2017:
http://eml.berkeley.edu/books/choice2.html

U.S. Census Bureau, "American Community Survey," undated. As of December 1, 2017:
https://www.census.gov/programs-surveys/acs/

U.S. Code, Title 10, Subtitle A, Part II—Personnel. As of December 1, 2017:
http://uscode.house.gov/browse/prelim@title10/subtitleA/part2&edition=prelim

U.S. Department of Defense, "Part One: Basic and Special Pay, Chapter 01: Creditable Service," DoD Financial Management Regulation, Vol. 7A, Chapter 1, April 2017. As of December 1, 2017:
http://comptroller.defense.gov/Portals/45/documents/fmr/Volume_07a.pdf